Fairmile 'D' MGB

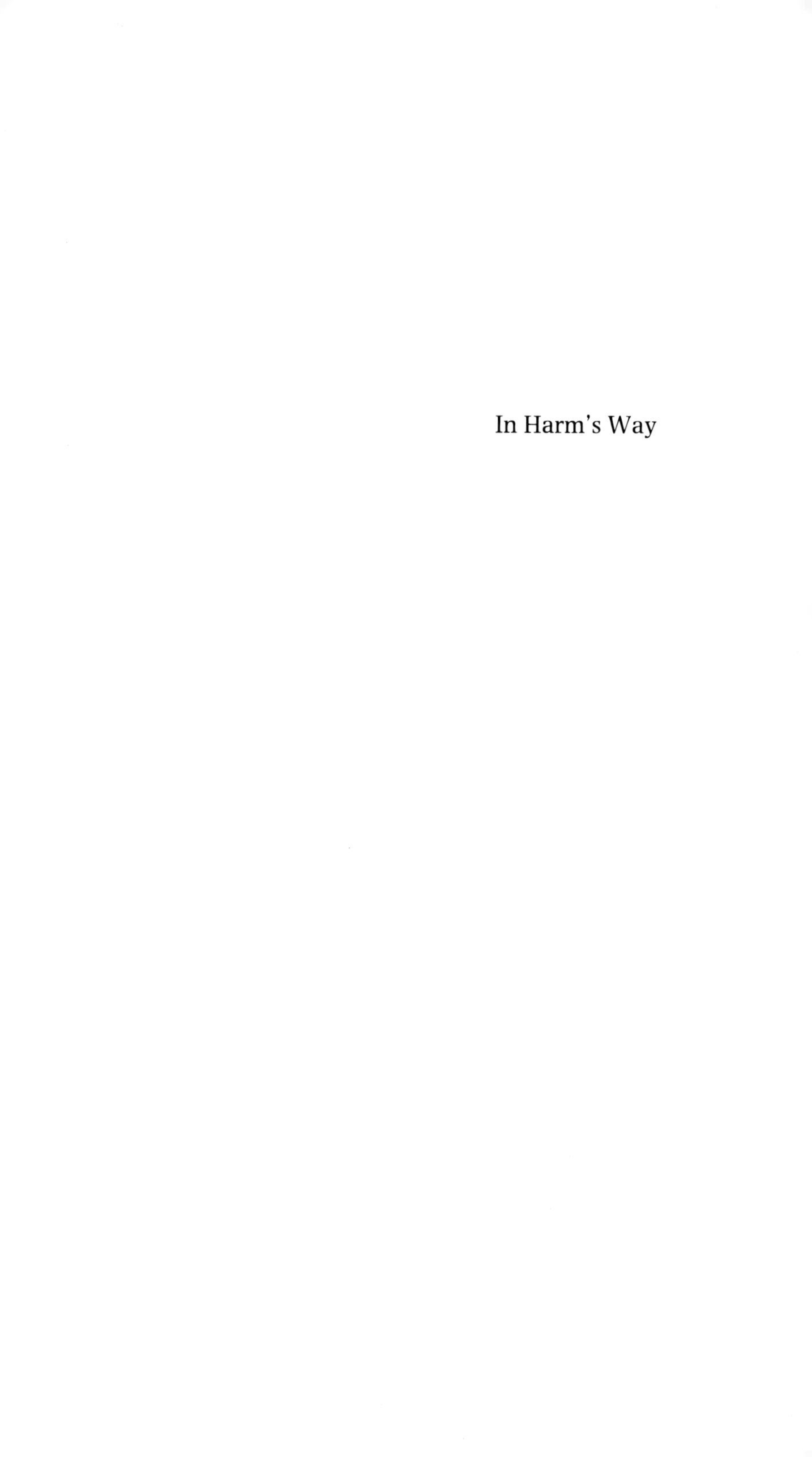

In Harm's Way

Geoffrey Hobday

In Harm's Way

A RNVR officer at war, 1940–44

'I wish to have no connection with any ship that does not sail fast for I intend to go in harm's way.'
John Paul Jones, 1747–92

Imperial War Museum

Published by the Imperial War Museum, Lambeth Road, London SE1 6HZ

Designed by Herbert and Mafalda Spencer
and printed in Great Britain by
Butler & Tanner Ltd, Frome, Somerset BA11 1NF

Maps by Line and Line
Endpaper drawing by Tony Garrett

Photographs (page numbers):
Imperial War Museum 22, 58, 60, 61, 89, 103, 108–109, 112, 122, 134–135, 141, 144–145, 149, 154, 159, 175, 186–187, 191, 201, 214, 215
Geoffrey Hobday Frontispiece, 15, 27, 39, 46, 52, 53, 55, 56, 63, 66, 67, 71, 73, 74, 77, 119, 139, 161, 171
Karl Müller 205, 207
Maurice Mountstephens 97, 113
Joan Thorpe 86
Rear-Admiral Sir Morgan Morgan-Giles 177
Jack Ivester Lloyd 193
Harold Garland 179

ISBN 0 901627 30 5

Frontispiece: The author in 1943

BUCKINGHAM PALACE.

As I spent two years plugging up and down the east coast in a converted World War I destroyer attempting to protect convoys from interference by E-Boats, among other hazards, I have good reason to appreciate what Geoffrey Hobday was getting up to in Coastal Forces.

The years inevitably take their toll of the eye-witnesses but histories will continue to be written about World War II long into the future and it will be from such personal firsthand accounts that historians will be able to give a flavour to the dry record of events.

Quite apart from the value of "In Harm's Way", as the material for historians, it is a remarkable and gripping story in its own right.

1985

This book is dedicated to all those fine men and women who served with Coastal Forces in the Royal Navy during the Second World War and to Helen and Courtenay who persuaded me to write this story.

Preface

This is the seventh volume in the personal experience series produced by the Imperial War Museum. It is, however, the first naval account we have published, and follows the highly successful secret diary of Dr Robert Hardie, *The Burma–Siam Railway* (1983). There could hardly be a greater contrast between the two accounts, for Dr Hardie recorded the terrible monotony and suffering of prisoners of war while Geoffrey Hobday's story is full of exciting and dramatic adventure. Yet both are faithful reflections of aspects of the Second World War and as such are valuable components of the historical record. Readers of this book will be left in no doubt that war, despite or perhaps because of its horrors, brings out remarkable qualities of courage, comradeship and enterprise. Commander Hobday also relates some of the amusing and unexpected incidents which helped to keep him and his colleagues sane even in the most difficult and dangerous circumstances.

The Museum continues to acquire letters, diaries, and other accounts relating to the various conflicts of the present century. These are held in our Department of Documents and provide source material for scholars and researchers the world over. I hope that readers of this account who have such documentary material themselves may be encouraged to donate it to us. The larger our archives become the better we, and future generations, can understand what modern warfare has meant to those who have been involved in it.

It remains for me to thank my colleagues who have worked to prepare the manuscript for publication. Once again the chief burden has fallen on Dr Christopher Dowling, who supervised all aspects of production, ably assisted by Mrs Janet Mihell. The result is a book which does credit to them and to the Museum.

Alan Borg, Director
Imperial War Museum

February 1985

Author's acknowledgements

The author wishes to thank the many people who have helped with the loan of photographs and documents and with the verification of incidents, dates and places. He is grateful to his sister-in-law, Kathleen Boreham, and to Norman Weddle for their continuing interest and encouragement and especially to his brother-in-law, Robert Dahl, who has acted as his literary agent and whose scholarly guidance has been invaluable.

Introduction

The Royal Navy's small ships, the motor torpedo boats, motor gunboats, steam gunboats and motor launches of Coastal Forces, fought seven hundred and eighty actions in European waters during the Second World War. They sank 70 German and Italian warships, including a cruiser and a U-boat, and about 140 merchant ships. The main purpose of these fragile, wooden-hulled, petrol-driven craft, some of which were capable of speeds of more than 40 knots, was to seek out and attack enemy coastal shipping but they were also used in many other roles. Operating mainly at night because they were vulnerable to air attack, they laid and swept mines, escorted troops and supplies, raided harbours and landed secret agents on remote beaches. Their opposite numbers were the formidable and elusive E-boats (*Schnellboote* or S-boats to the Germans), whose low silhouette, sea-keeping qualities and Daimler-Benz diesel engines gave them an advantage over the more inflammable and less reliable British boats.

The 'Mosquito Fleet', as Churchill dubbed it, was manned almost entirely by Royal Naval Volunteer Reserve officers and ratings, many of them dashing individualists who had been yachtsmen or racing car and motorboat enthusiasts before the war. Like the so-called 'long-haired boys' of RAF Fighter Command, the 'amateur' sailors of the RNVR tended to be self-reliant, unconventional, insouciant and not over-deferential to authority. Geoffrey Hobday, a New Zealander, who managed to join Coastal Forces despite at thirty-three being over-age, was one of this special breed. *In Harm's Way* is an exciting and fast-moving personal account of small boat warfare in the theatres in which Coastal Forces were most active, the North Sea, the English Channel and the Mediterranean. As captain of an 'A' and later a 'B' class motor launch based on Great Yarmouth, he was involved in the bitter struggle to keep the vital East Coast convoy routes open in the face of relentless attacks by E-boats and the Luftwaffe. From 1943 to 1944 he commanded a Fairmile 'D' motor gunboat in the Mediterranean, which, with its relatively calm waters, numerous islands and abundant coastal shipping, offered considerable scope for the employment of small craft. Here, in addition to E-boats transferred from the English Channel by means of the German and French canal systems, Coastal Forces encountered the fast but lightly armed Italian MAS and MS boats, powered by the excellent Isotta-Franchini engine. Geoffrey Hobday's MGB 643, which was attached to the 19th MGB Flotilla, operated off North Africa, in the Straits of Messina, the Tyrrhenian Sea and the Aegean, emerging unscathed from the ill-fated Dodecanese campaign. Later MGB 643 played a prominent part in the piratical warfare which the Royal Navy waged in the Adriatic in support of Tito's partisans. During the last few months of his service with Coastal Forces, Geoffrey Hobday commanded a 'D' boat in the North Sea, where, after an action in September 1944, he rescued the noted E-boat leader, Karl Müller.

When opposing motor torpedo boats came in contact spectacular battles sometimes took place, fought at bewildering speed in flying spray amid a hail of tracer bullets. Survival depended upon split-second judgement, ice-cool nerve and expert marksmanship, especially as accidental sinkings were not uncommon. Yet, as Geoffrey Hobday's account reminds us, such encounters were exceptional and were interspersed with long hours of monotonous escort duty and uneventful patrols. This is borne out in the record of Lieutenant-Commander Robert Hichens, who became something of a legend in Coastal Forces. When he was killed by a stray burst of fire on 12 April 1943 Hichens had completed 148 operations, but these included only fourteen actions.

Geoffrey Hobday did not keep a diary during the war (private journals were expressly forbidden in the navy) and inevitably there were points in his narrative that needed to be checked or filled out. In the task of preparing the manuscript for publication I have been greatly helped by the author's brother-in-law, Robert Dahl, formerly Head Master of Wrekin College, and by my colleagues Janet Mihell, Roderick Suddaby, Captain Arthur Wheeler RN and Commander Ronald Fisher MBE RN. I am also indebted (in some cases for permission to reproduce photographs) to Geoffrey Hudson, Lord Strathcona, Rear-Admiral Sir Morgan Morgan-Giles DSO, OBE, GM, M R D Foot, Mrs Joan Thorpe, Lady Craven, Douglas Hunt DSC (Honorary Secretary of the MTB Officers' Association), Claude Holloway DSC (Honorary Secretary of the Mediterranean MTB Officers' Association), Len Bridge (Secretary of the Coastal Forces Veterans Association), Karl Müller, Maurice Mountstephens DSC, Jack Ivester Lloyd DSC, Mrs Pauline Cutler, Harold Garland and the staff of the Naval Historical Branch, Ministry of Defence. Finally thanks are due to David Cobb, himself a veteran of Coastal Forces, for the painting reproduced on the dust jacket, which was specially commissioned for the book.

Christopher Dowling
Keeper of the Department of Museum Services

Maps:

1

I shall never forget that wonderful morning of Friday 10 May 1940. As I walked out of the New Zealand Government Offices in the Strand into a sparkling London spring day I felt a great glow of elation which seemed to spread from the document in my pocket that the High Commissioner, Mr Jordan,[1] had just given me. This instructed the managing director of my employers, the large paper firm Albert E Reed and Company (now Reed International), that I was to be released forthwith from the 'reserved occupation' status they had clamped on me. I had been accepted for the Royal Naval Volunteer Supplementary Reserve some months before the outbreak of the Second World War. The reserved occupation order forced me to resign from the RNVSR and also effectively prevented me from joining any of the other fighting services.

Two weeks after my meeting with Mr Jordan I attended an Admiralty Selection Board at Queen Anne's Mansions for a commission in the Royal Naval Volunteer Reserve. The board that interviewed me consisted of a rear-admiral and three captains from the executive, engineering and paymaster branches respectively. The interview had been preceded a few days earlier by a medical. The board were very efficient and bombarded me with questions designed to test my grasp of seamanship, engineering and navigation as well as my general knowledge. The way that the candidate coped with the questions enabled the board to assess his self-confidence and leadership potential. I was successful in passing this searching examination but it was a good three months before there was a vacancy for me at HMS *King Alfred*, the new RNVR training centre for officers which was situated on the Sussex coast at Hove. Such naval shore establishments were often referred to as 'stone frigates'.

[1] Later the Rt Hon Sir William Jordan (1879–1959), High Commissioner for New Zealand, 1936–1951.

I joined *King Alfred* with the rank of sub-lieutenant in August 1940 and was measured for my uniform almost immediately; all the well-known London naval and military tailors had rented shops nearby. I can remember how inordinately proud we novices were of that single thin wavy ring of gold braid which signified our rank. At the same time, for the first few days we were very self-conscious whenever we wore our beautiful new uniforms in public. I even felt uneasy when I walked down the street lest an emergency arise and someone say, 'Stand back. Here's a naval officer. He'll know what to do!'

Our training programme was highly concentrated. The urgency of the situation was so great that the course had to be completed in a mere five weeks. Staff and trainees worked like Trojans to achieve the maximum in that desperately short time, but we could only touch the fringes of the broad spectrum of knowledge which the duties of a naval officer demanded. No boats were available and only the British genius for improvisation made it possible to carry out any practical training. We used Wall's ice cream vendors' box tricycles to practise fleet manoeuvres, pedalling them around at high speed and with great *élan* to 'steam' in line ahead, line abreast and other formations. Realism was added to our make-believe ships by miniature masts fitted with halyards and signal flags, so that we could make and receive the various manoeuvring orders. We rang our bicycle bells, which substituted for ships' sirens, on every possible occasion to signal, 'I am turning to port', 'I am turning to starboard', 'my engines are going astern', and so on. At times it was all so hilarious that we nearly collapsed with laughter; but I suspect that we learnt these manoeuvres far more quickly on our tricycles than we could have done in ships.

I found that I enjoyed two advantages over many of my classmates. First I was an experienced yachtsman with a good knowledge of chart work and coastal navigation and, secondly, thanks to my New Zealand background, I was a trained soldier. For many years military training had been compulsory in New Zealand for all young men between the

Passing-out class, HMS *King Alfred*, September 1940.
Sub-Lieutenant GM Hobday is second from the right in the front row.

ages of sixteen and twenty-one. It was carried out on a part-time basis but was very thorough. I served in an infantry battalion for two years and then in a battery of field artillery.

As we neared the end of our course, we were interviewed individually by the senior training officer, Commander Head, and told what our first appointment would be. When he informed me I was to be assigned to the RAF as a liaison officer, I pleaded with him to try to find me a sea-going posting instead. He was always poised and courteous – he would have made a very good career diplomat – and he had a keen sense of humour, which he was able to exercise when he sent for me again the following day. 'You are now being appointed to HMS *Maron* as a watch-keeping officer,' he said. 'She's a 7,000-ton cargo ship and is being armed and converted to become an ocean boarding vessel for Atlantic patrol. She's diesel-engined and can stay at sea for forty days or more without refuelling. Now, Hobday, you couldn't get anything more sea-going than that; and you might even encounter one of the Jerry surface raiders like the *Scheer* or the *Hipper*.'

A few days later I was given my appointment papers and sent on seven days' leave before reporting to the *Maron*. I went straight to Sutton, a suburb of London, where my English-born wife Helen was teaching at the Sutton Girls' High School and was in digs with our seven-year-old son, Courtenay. The blitz was in full swing and they had had some close calls with bombs exploding nearby. That week together meant a great deal to us. In those dangerous times we had no idea when or if we would see each other again.

The *Maron* was berthed in Liverpool Docks when I reported on board to the captain, a thick-set florid man in his late forties or early fifties. He carried the rank of commander RNR and was wearing First World War medal ribbons. He wasn't very friendly and openly showed his disappointment when he learned of my lack of training and of deep sea experience. However, he cheered up a little when I was able to convince him that I had had a thorough grounding in artillery and small arms. He then delineated my duties, which were manifold. My military experience made me a natural for selection as captain of the quarters for our after 6-inch gun, and as a boarding and prize officer. In addition I was to be an officer of the watch and a divisional officer – this is a kind of welfare officer, whose job is to help ratings with any personal problems.

I was told to report our discussion to the first lieutenant, a likeable RNR lieutenant-commander, who introduced me to several of his colleagues, all RNR and all professional merchant service officers. I was to be the only 'amateur' officer appointed to the ship. The first lieutenant explained that it would be several weeks before the *Maron* would be ready for her naval service, and the alterations and additions required were legion. She was not yet in commission and meanwhile carried only a skeleton crew. While the accommodation on board was being altered the ship's company were living ashore in digs. The *Maron* was an ex-Blue Funnel Line ship, heavily built to withstand typhoon weather in the Far East. Her armament would comprise two old 6-inch guns, to be mounted forward and aft, plus four old Hotchkiss machine guns for anti-aircraft work. Our full complement of officers and men would total about one hundred and forty, which was double that of her peacetime crew. This great increase in manning was essential to provide for guns' crews, extra wartime lookouts and armed guard boarding parties, who would be needed to take over any ships that we might capture as prizes on our patrols.

Here I should explain some of the naval terminology of the time. There were three categories of commissioned officers – Royal Navy, Royal Naval Reserve and Royal Naval Volunteer Reserve. Seniority was in that order. RN were regular officers of the Royal Navy. RNR ('Rockies') were professional merchant marine officers serving in wartime in the navy, and RNVR ('Wavy Navy') were ex-civilians and not professional seamen. A saying popular at the time explained the differences in this way. 'The RNVR are gentlemen trying to be sailors and the RNR are sailors trying to be gentlemen; but the RN are neither, trying to be both.' Another common designation was HO (hostilities only) applied to men serving in the navy for the duration of the war. T124 ratings were merchant seamen in navy uniform, carrying naval rank, and were under naval command and discipline but paid at the much higher rate of merchant seamen; in addition, they often received special danger money. In ships such as the *Maron* with mixed crews of naval and T124 men who shared the same dangers and hardships together, this pay anomaly could engender a lot of ill-feeling. As a watch-keeping officer I was in command of the ship when on the bridge, yet an ordinary T124 rating's pay was twice as much as my own.

It seemed a long time before the *Maron* was ready for sea. At last the accommodation was finished, the guns were mounted, the Carley life-saving rafts were fitted, large naval signal lamps were installed on the bridge, we were equipped with gun telephones and recognition signals, and thousands of empty forty-gallon oil drums were stored in the holds. The object of the drums was to provide a degree of flotation if we were holed by shells, torpedoes or mines.

We had a lot of trouble with the after 6-inch gun mounting, which had to be secured to the superstructure instead of directly to the main deck, as was the normal practice. In spite of all efforts to brace the superstructure, a local earthquake of astounding magnitude occurred every time the gun was fired. Both of our 6-inch guns had interesting, but not very reassuring, history sheets. One of them had been condemned as unfit for further service in 1906 and the other in 1907.

After completing trials, we sailed for the Clyde where a big Atlantic convoy was being assembled off Greenock. On 6 December 1940, while awaiting sailing orders, we experienced a freak storm. The captain and one watch were ashore on short leave; by late afternoon it was blowing too hard for any of the liberty boats to get them back on board. It was a terrible night, pitch black with no lights ashore or afloat because of blackout regulations; and driving rain and sleet added to the obscurity. At 4am when I took over the watch on the bridge the wind was gusting so violently that it was impossible to stand except by holding on tight to a handrail. At 4.30am, with the wind still backing and increasing, I called out the first lieutenant, who immediately put power on the steering and placed the engine room and deck crews on stand-by.

The second of our two main anchors had been dropped three hours before and more cable had been veered to prevent dragging. We were now compelled to steam ahead on both main engines to relieve the terrific strain on our anchors and cables. By 5.30am the barometer had dropped to the lowest reading I had ever seen and the anemometer recorded gusts of up to 100 knots. The ship held all right, and we were mightily glad of the heavy, oversize anchors and cables which had been designed for typhoon condi-

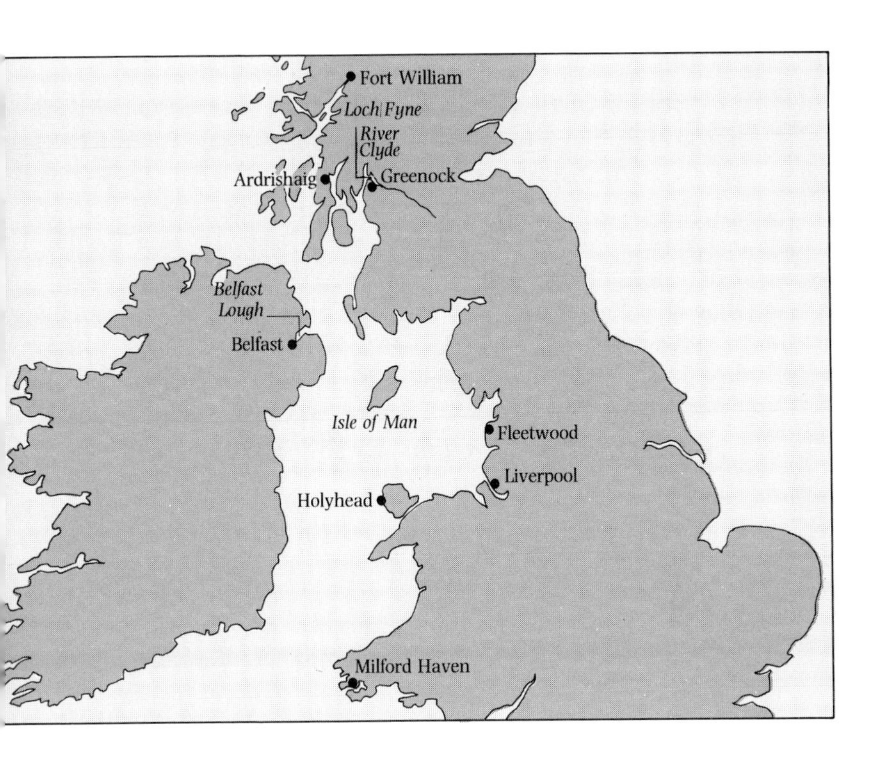

tions. Many other ships dragged their anchors or parted their cables, and we had some narrow escapes as the runaways surged by out of control. In the end the situation became so dangerous that we had to ignore the blackout and use our big signalling lamps as searchlights, flashing them on every now and then. One of the runaways scraped and gouged her erratic way all along our port side and we were lucky not to be severely damaged.

When dawn came we found that more than a dozen ships had been blown aground, including a 30,000-ton Cunarder. The storm delayed our departure for nearly a fortnight and when eventually we did sail there were several ships which were not yet seaworthy enough to join us. It was about mid-December 1940 when we steamed out of the Firth of Clyde with the convoy, which we had been ordered to accompany for a few days as an additional escort before heading for our patrol area. We weren't much of an escort. We were not equipped with radar or Asdic, nor did we carry any depth charges, so there was nothing we could do against U-boats, and our armament was hopelessly inadequate to cope with surface raiders or aircraft.

In winter, even in a luxury liner in peacetime, the North Atlantic is a cold, grey and stormy sea. It soon became clear that this trip would be no picnic. When the *Maron* was converted from cargo ship to warship she had to be heavily ballasted to bring her down to her marks. Unfortunately, the ballast, comprising pig-iron and concrete, was stowed right down along her keel instead of being properly distributed. This resulted in the ship becoming overstable, with a jerky and viciously quick roll and recovery. In fact our righting movement was so pronounced that under certain conditions a pendulum-like effect would build up to an alarming extent. In a heavy beam sea or swell the *Maron* would roll between 35 and 40 degrees each way, and on a few occasions the reading on the inclinometer exceeded this. Nothing can be more nauseating and tiring than this sort of motion, and most of our crew were seasick. But seasick or not, except in the most extreme cases, they had to keep their watches and carry out their duties. No one on board had ever experienced anything quite like this before. Once or twice during some of the worst of the rollings, I had the extraordinary experience of being hurled across the bridge to find myself standing on the doors of the signal flag lockers, which are normally vertical, with my hand stretched out sideways to the deck to steady myself.

These hardships were worst for deck officers and seamen, as effective look-out could be maintained only in the open, with little protection against wind, rain and cold. We suffered a lot of damage and breakages. The thin steel plate of some of the lifeboats was badly dented when the gripes securing them gave way. The concrete blocks which protected parts of the bridge and the wireless room came adrift and collapsed. Down below, we were little better off. It was very difficult to get enough sleep. After being thrown out of my bunk a few times, I developed a method of locking myself in it by lying on my stomach and bracing my elbows and legs between the inner bulkhead and the outer bunk-board.

In the first few days, nearly all our crockery was smashed and it was often quite impossible for the cooks and stewards to prepare meals, let alone cook them. We had to live on quite a lot of hard tack. The noise down below was terrible. There had been some movement in the oil drum stowage, and the crashing and banging of the empty steel drums made a tremendous din, which had a most wearing effect on the nerves.

Weather conditions started to ease soon after we left the convoy and they continued to improve as we steamed southwards to take up position in our patrol area. A large part of the western North Atlantic was patrolled by a variety of vessels – armed merchant cruisers, ocean boarding vessels like the *Maron*, corvettes and armed trawlers. These patrol vessels had a two-fold purpose, which was partly to notify the Admiralty of any sightings of German surface raiders or U-boats, in which case the object was to get a wireless message off with position and description before being sunk, and partly to intercept any enemy merchant ships, including those taken over from the occupied countries. These were to be seized and brought to the nearest British port under armed guard.

RAF Coastal Command did not have suitable aircraft to carry out really long reconnaissance patrols, and the area they could cover was further limited by the lack of bases between Britain and Gibraltar. Our Sunderland flying-boats did some fine work, but they could hardly be described as fighting aircraft. On the other hand, the Germans had a splendid long-range machine, the Focke-Wulf Condor, which, shuttling between bases in France and Norway, was able to patrol large expanses of the North Atlantic.

Things started to happen almost immediately we reached our patrol area on Christmas Eve 1940. We received coded messages from the Admiralty plotting a number of U-boats thought to be nearby, and then a further signal stating that the German heavy cruiser the *Hipper* was also believed

HMS *Maron*, ocean boarding vessel.

to be in our vicinity. We were all very much on the alert but Christmas Eve passed without incident. Around midnight, just after I had taken over my watch on the bridge, another Admiralty coded message arrived, advising us that a British naval unit would be intersecting our course at about 1am. We barely had time to assimilate this before we sighted an aircraft carrier, a cruiser and a number of destroyers streaking across our bows in an easterly direction. Twenty minutes later the western horizon lit up with flashes and we could hear heavy and frequent gunfire. This went on for about half an hour and then ceased as suddenly as it had started. By this time we were closed up at our action stations wondering what would happen next.

We assumed that some British ship had engaged the *Hipper*, but were greatly puzzled at having previously seen a powerful British force speeding away from the scene of action. We soon had confirmation from the Admiralty that the gunfire had come from a skirmish between the *Hipper* and HMS *Bonaventure*, a 6-inch cruiser. The *Bonaventure* was no match for the *Hipper*'s 9-inch guns but managed to drive her opponent off, both ships suffering some damage and casualties.

It was several days before I learnt the reason for the mysterious action of the aircraft carrier and her escorts. The aircraft carrier was an old First World War converted battle cruiser, the *Furious*, which was being used as a cargo ship. Aboard was probably one of the more important war-time cargoes carried up to that time, consisting of a large number of aircraft in cases, together with aero-engines and other spares. This freight was to be landed in Takoradi in West Africa for assembly, and the aircraft were then to be flown to Egypt. The Admiralty decision, quite rightly, was that it was far more important to get all these aircraft safely to port than to risk their destruction by the *Hipper*. The *Bonaventure* was therefore detached to deal with the *Hipper* on her own, leaving another 6-inch cruiser and the destroyers to continue their escort duties with the *Furious*.

There were no further alarms that night, but at first light a sizeable Dutch oil tanker loomed up quite close to us. We trained our guns on her and ordered her to stop for investigation. Her replies were slow and not very satisfactory, so we were greatly surprised when our captain allowed her to proceed without sending an armed boarding party to carry out a thorough search on board her. Nearly all of us were convinced that she was the refuelling ship for the German surface raiders, and that she was about to make a rendezvous with the *Hipper* or had recently done so. In any case, this was just the sort of ex-neutral ship that would now be under German control and should be taken in prize by us. Our captain had shown extreme nervous tension ever since we had sailed from the Clyde, and this quite irrational decision further diminished our confidence in him.

At 8.30am we sighted smoke on the horizon and increased speed to close it. Two hours later we were within range of a vessel with huge French tricolours painted on her sides. We trained our guns and made a flag hoist in merchant code, ordering her to stop. This she completely disregarded, and I was instructed to put a shot across her bows. Although I knew that almost anything could happen when I fired my clapped-out 6-inch gun, I was still quite unprepared for what occurred. With elevation correctly set for the range, we were horrified when we saw the shell slowly leave the muzzle, turn end over end several times in the air and plunge into the sea only a quarter of a mile away. This firing was accompanied by the recoil-generated 'earthquake' and the crashing sound of breakages down below. Owing to the worn-out state of the barrel, tongues of flame from the propellant leaked past the shell and momentarily enveloped all of us on the gun platform.

Nothing daunted, after sponging out the barrel to ensure that no burning particles of propellant remained, we prepared to fire our second round. This time I gave the gun much greater elevation than the range warranted, and hoped for the best. We were very lucky with this shot, which was perfect for range though somewhat too close to the French ship's bows for comfort. The Frenchman didn't disregard this one and stopped smartly. We were so pleased with our efforts that we cheered and slapped each other on the back. Then the gun telephone sounded with a message that I was to report immediately to the captain.

As I made my way to the bridge I thought, 'He's going to tear one hell of a strip off me for putting that shot too close,' but the summons turned out to be about something quite different. 'Hobday,' the captain said, 'you are to board that ship and take her to Gibraltar under armed guard. The seaboat is being made ready and so are the necessary charts and other papers. You've got ten minutes to grab your gear before reporting back here.'

This was indeed a bolt from the blue. Gibraltar was 500 miles away and my knowledge of deep sea and astro-navigation was nil. I didn't even know how to use a sextant. Despite the fact that I was the least experienced of the eleven deck officers we carried, I did not protest my ignorance. Regardless of what problems I might have to face, I thought that seizing and taking command of a prize was rather an exciting prospect, and it would certainly enrich my experience. Also, I was glad to have the opportunity of getting away from the *Maron*, if only for a short period.

All these proceedings were highly irregular and the captain had no right to send me away in the manner he did. There was only one copy of the Prize Manual aboard the

Maron; this was being passed around by the officers in order of seniority and had not yet reached me. When I read it later in Gibraltar, I found that, among other things, it laid down quite clearly that before the commencement of each patrol a roster must be prepared so that the armed guard officers would know beforehand the order in which they would be required. Furthermore, it clearly stated that, when ordered away in the seaboat, the armed guard must take his own time so he could satisfy himself that he was provided with all the charts, equipment and information that he needed. Had I known this at the time I could have saved myself a lot of heartache.

When I returned to the bridge, I was handed a rolled-up parcel containing my orders, charts and recognition signals. I was also given a .45 service revolver with ammunition, and £50 advance of pay and expenses. After I had strapped on the revolver, I was sent to the seaboat, where the boarding officer, his signal rating and the armed guard were waiting for me. I noted that our provisions were already stowed in the boat, jumped in and gave the order to lower away and slip.

My detachment, equipped with rifles and bayonets, numbered four – Leading Seaman Davies and Ordinary Seamen Searle, Brown and Cook. Only one of these had been to sea before and they were even greener than I was. Our reception when we boarded the *Joseph Duhamel* and told the crew we were taking their ship to Gibraltar could hardly be described as friendly, and some of the insults hurled at us were quite unprintable. If we hadn't immediately made a firm show of force, there could have been real trouble then and there.

We found that the *Joseph Duhamel* was a large ocean-going trawler of over 1,000 tons. She was a single screw, coal-fired steamer with a crew of fifty-three, carrying a cargo of 500 tons of salted cod. The Vichy French government had ordered her to sail from the island of St Pierre in the Gulf of St Lawrence, and she was on her way to another French possession, Dakar in Senegal, when we intercepted her. Our first move was to put the wireless out of action and to hold the captain and mate as hostages on the bridge, while we searched the vessel from stem to stern and familiarised ourselves with her layout and equipment. All this was done with the utmost speed, and soon the boarding officer and signal rating were returning to the *Maron* in the seaboat.

The moment they were clear, I rang down 'Full Ahead' and set course for Gibraltar. It was a matter of urgency now that we made the bridge into a citadel. There were only two points of access, both by ladder, and I was able to secure our position by placing an armed sentry at the top of each. As we also commanded an all-round view of the whole of the upper deck, we would have good warning if any attempt was made to rush our position.

Bringing a prize to port is a most unenviable task, requiring unremitting vigilance and making heavy demands on physical and mental endurance. I knew that our captives would be trying to work out some scheme to overpower us and would not hesitate to kill if necessary. If they failed, they might resort to sabotage. I couldn't be certain that there were no concealed firearms in their possession, but I could be sure that every seaman would have a sheath knife and it is obligatory for every ship to carry a number of junk-axes. We were outnumbered by fifty-three to five, and could feel the French crew's hostility as though it was a tangible thing. At this stage of the war the French probably hated the British even more than they hated the Germans. Deep down they suffered from a wounding sense of shame over the capitulation of France, and in an effort to sublimate this they blamed us for 'deserting' them by evacuating our army at Dunkirk. Our greatest problem, with Gibraltar nearly three days away, would be to remain continuously on the alert throughout that period – and after our experiences in the *Maron* we were already short of sleep.

One of the epithets the French had used when we boarded the *Joseph Duhamel* was 'damned pirates'. It now struck me that technically this was quite true, as we were not at war with France. The Royal Navy had been ordered

The *Joseph Duhamel*, looking forward, February 1941.

to seize all French ships found on the high seas, partly as a retaliation for the seizure of British ships in French ports at the time of the surrender of France, and partly to prevent any possibility of their falling into German hands. In accordance with my orders, I was flying the White Ensign at the mast-head, with the French tricolour immediately below, showing quite clearly that, according to international law, I was engaged in an act of piracy. Piracy can be punishable by death, so I devoutly hoped that we were not intercepted by any French warships. I didn't relish the possibility of facing a firing squad, or worse still, being dragged to the guillotine. Although a substantial part of the French fleet in North Africa had been dealt with by the Royal Navy in July 1940, there were some French units still at large based on their West African and West Indian possessions.

I found that I had quite a few problems on my hands.

The vital navigational parcel, handed to me on the *Maron*, proved to be seriously incomplete. The only chart was of the Straits of Gibraltar, and the recognition signals form hadn't been filled in. Without the proper signals we were placed in a very dangerous position. A British ship would probably open fire on us if we didn't give the appropriate response, and to approach the great fortress of Gibraltar without answering a challenge correctly would be to court disaster, particularly at night. I then discovered that the powerful Aldis signalling lamp, which should have been left with us, had been taken away in the seaboat when it returned to the *Maron*.

But, in spite of these anxieties, I had a tremendous feeling of exhilaration. This was my first command, and a command under difficult circumstances, thrust upon me within weeks of joining the navy. I decided that I must try to convert the hostility of the French crew to acquiescence or preferably to cooperation, so I went to work on Captain Lecoeur and his son André, who was the mate. Both of them were still being held on the bridge. André could speak English fairly fluently and I had been talking with him a good deal. From this an easy relationship had developed, and I sensed that he could become a friend. I managed to convince them that they had nothing to fear, provided we remained unmolested and were able to bring the ship safely to Gibraltar, otherwise the consequences for them would be very serious indeed. I told them (quite untruthfully) that our forces in the area had been instructed to keep us under surveillance, and that they would react sharply and ruthlessly if they suspected any trouble on board.

'On the other hand,' I said, 'if everything goes smoothly, I will see to it that on arrival in Gibraltar you are all treated generously. Everyone will be given an absolutely free choice between joining up with de Gaulle and the Free French forces or being repatriated to France. I give you my word as a British naval officer on this, and in return I would appreciate a promise of your cooperation.' I was delighted when, after a brief consultation together in French, they readily agreed. I released them shortly afterwards, enjoining them to inform all members of the crew, at the same time warning them that under no circumstances was anyone to approach the bridge without good reason, otherwise we would open fire. Perversely, I almost hoped some hothead might try.

Towards midnight we heard scuffling near the bottom of the forward bridge ladder, and I immediately let fly with a round from my .45 pistol, deliberately aiming high. There followed the sound of feet running along the deck, which petered out into an eerie silence. We were not disturbed again in this way. What it was all about I never knew, but the incident had proved that we were not to be taken by surprise and that we would shoot.

Another piece of luck presented itself next morning in the form of an RAF Sunderland flying-boat whose pilot challenged us by Aldis lamp. On receiving no response, he came in quite close and had a good look at us. I was tremendously cheered by this encounter and we waved madly from the bridge. I was certain the French would now be quite convinced of my surveillance story. At the sound of the Sunderland's engines, a lot of the crew had come up from below, and had seen the signalling and the RAF roundels on the flying-boat's wings and hull.

Shortly afterwards, André hailed me from the deck and I invited him onto the bridge. His news was good, and, as I had hoped, the Sunderland visit had dispelled any doubt that what I had told him was true, and that the men could rely on the promises I had made. André did not think we would have any trouble, although most of the crew would want to return to France and only about a third of them, including himself, would opt to join the Free French. I decided to maintain strict vigilance and discipline, and not to relax them until we were safely in port.

Ever since setting course for Gibraltar, I had been wrestling on and off with various navigational problems. The log was quite inaccurate and the compass was suspect, so that I could not be sure of the course made good nor the distance run. I dared not seek help from André or his father: they could have fed me false information and, worse still,

I would have disclosed my total lack of experience. Furthermore, I knew of no way to plot our progress without a chart showing the area from where I had taken over the *Joseph Duhamel*; but all I had was the chart of the Straits of Gibraltar. However, I did have our departure position and the magnetic and true courses to steer for Gibraltar, which had been given to me by the navigating officer as I left the *Maron*. My true course was 080 degrees, which meant there would be only a slight change of latitude as we ran our distance. It then dawned on me that I could use the Gibraltar chart to devise a table by ruling our course on it and measuring off the resulting changes in latitude and longitude for any given distance. This worked perfectly – what I had done was to invent a traverse table. These are part of every standard book of nautical tables and have been in existence for a very long time, but I didn't learn that until later.

It now remained to find out what course to steer by our compass to effect a true course of 080, and to check what speed we were making. I worked out the approximate compass error by taking a bearing on the Pole Star, which I knew was never more than 2 degrees away from true north, and I judged our speed at about nine knots, which André confirmed. I experienced a terrific feeling of elation.

I could now concentrate on the only remaining problem – entry into the narrows of Gibraltar in the dark without a signalling lamp or recognition signals. I should mention that my written orders imposed absolute radio silence so it was impossible for us to communicate. From my new navigational data I had already worked out that we should arrive between eight and nine o'clock the following night. I came to the conclusion that the safest bet would be to make as open an approach as possible, with navigation lights on and with the bridge chart room fully lit up. I would then heave to and use the main switch on the chart room lights to flash the SOS distress message.

Next morning I sent for André and told him to put the crew to work and to get them and the ship spruced up. I thought it wise to keep them as fully occupied as possible.

I also told André that a ship and crew that were smart and clean would make a favourable impression on the authorities at Gibraltar and would help me in my efforts to secure the best possible treatment for them. He agreed with this and retorted, 'It will help our French pride too when we meet the British there.'

There was great excitement on the bridge when at about 11.30am land was sighted on the starboard bow. As I had fervently hoped, this proved to be Cape Spartel, whose lighthouse near Tangier marked the entrance to the straits on the African side. By the time the cape was abeam I had completed a running fix, which put our position just a mile north of the course that had been marked for us on the *Maron* chart. Our improvised navigation had worked out so well that we had made an almost perfect landfall, giving me a thrill I shall never forget. This was my first landfall in my first command.

We carried on up the straits until 8.30pm, when we were some three miles south-west of Gibraltar's Europa Point. Here I stopped ship as signals flashed to us from the rock. We replied by signalling with our chartroom main switch. A powerful searchlight was trained on us and we were approached and hailed by a patrol vessel. I reported briefly and was told to proceed into Gibraltar Bay, where a pilot came aboard and brought us up in the anchorage off the Waterport. Shortly afterwards, we were boarded by a naval party from Contraband Control, who inspected the ship's and crew's papers and took details of our cargo. We had made it, in spite of many difficulties.

Within two or three days of my departure from the *Maron* she intercepted another French ship, which, after being taken in prize, disappeared. The armed guard party were discovered in Dakar some fifteen months later, when French West Africa came over to the Allies. They had been overpowered and put into solitary confinement under the most abominable conditions. This could well have been our fate also.

Once Contraband Control had departed, we lost no time in getting some sleep. I had had a total of four hours' sleep over a period of three days and nights, and the ratings had had little more. They had all behaved splendidly and I later commended them in writing to the captain of the *Maron*. We were all recommended for hard-lying money, and we felt that we had earned it.

2

Next morning I went to see the RN commander who was the executive officer of the Contraband Control base. My reception was a frosty one and the commander demanded a full explanation of my failure to carry out proper recognition procedures the previous night. 'You caused us a lot of trouble,' he said, 'and you could have been blown out of the water. The 9.2-inch gun battery was loaded and trained on you.' 'I couldn't do anything about it, sir,' I apologised, 'because I had no recognition signals and no signalling lamp.' I then showed him my blank signals form and told him something of my experiences since leaving the *Maron*. When I finished, he was almost speechless with astonishment and from that moment appeared to have my interests at heart. He kindly arranged for me to dine as a guest in the Gibraltar artillery mess, where I was welcomed as a fellow gunner, and to have drinks with Captain Egerton of the *Bonaventure*, which was in Gibraltar for repairs after her action with the *Hipper*.

My leave ashore was terminated abruptly and unexpectedly on the afternoon of 1 January 1941. A telephone call instructed me to report at once to Contraband Control, equipped for armed guard duties on a large French prize about to enter Gibraltar Bay. At the base, a surprisingly large detachment was already being assembled and we were quickly mustered and briefed. Our armed guard was made up of fourteen Royal Marines, with a sergeant and corporal, and sixteen naval ratings with two petty officers. A RNR lieutenant was in command and I was his Number

Two. All were regulars, well armed in full battle order, and they looked as though they could cope with anything.

The prize was the 16,000-ton Messageries Maritimes passenger liner, *Chantilly*, which had been intercepted off Dakar by a British destroyer under most unfortunate circumstances. One of the destroyer's machine guns had accidentally discharged, killing two women passengers on deck and wounding many others. The *Chantilly* had some four hundred and fifty passengers, many of whom were French naval and marine officers and men. Their families accounted for most of the rest. We were warned that there was bitter anti-British feeling on board and that the atmosphere was explosive. The French officers still carried their side arms but it was not known what other weapons were on board. As it was anticipated that our task would be an arduous one, arrangements were being made for a second armed guard detachment to alternate with us.

After this briefing, we were taken out in a liberty boat to board the *Chantilly* as she anchored in the bay. We relieved the destroyer's armed guard, who looked exhausted, and they confirmed that they had had a very difficult passage. The next few days were among the most unpleasant I have ever had to endure. Although the passengers and crew had been informed that the firing of the machine gun was nothing more than a most deplorable accident, and full apologies and compensation had been offered, their attitude towards us was a mixture of almost incandescent hatred and icy contempt. None of them would speak to us, and if spoken to they would deliberately turn their backs. But they made quite sure that we should know their opinion of us by expressing it loudly whenever we passed by. We could get no cooperation or understanding from them at all.

Soon after boarding the *Chantilly* we tried to arrange for the two bodies to be taken ashore discreetly at night, but the French would have none of this. On the contrary, they insisted on holding a service on deck round the coffins, which were draped with the French tricolour and were ready for lowering over the side to a waiting boat. The whole emotionally charged drama was played out in a manner calculated to intensify their bitterness. After this ceremony some of them spat in my face. It was significant that out of the *Chantilly's* total complement of over eight hundred, only three volunteered for the Free French.

The days dragged by until on 7 January my stint of armed guard duties aboard the *Chantilly* was terminated as suddenly as it had begun. That morning a picket boat came alongside that hateful ship, bringing a relief officer to take over from me. There was some very interesting news for me at the Contraband Control base. The *Joseph Duhamel* had to be sailed in convoy for England at noon on the following day. There was no merchant master available to take command of her, and the Commander-in-Chief's office had been in touch with the base to find out if I would be prepared to volunteer for the job. It was essential that the *Joseph Duhamel* should have a properly appointed master, but the navy had no powers to order me to take command.

I liked the idea of this new assignment and volunteered at once. I had had more than enough of the *Chantilly* and I had no desire to rejoin the *Maron*, which was not a happy ship. But to be given an appointment as a merchant master appealed to me enormously. Everyone seemed pleased about my decision, but Contraband Control were so overworked at the time that they couldn't give me any direct assistance. They did, however, allow me *carte blanche* to go ahead and make what arrangements I could.

It didn't seem possible to me to get the ship ready for a long voyage in a mere twenty-four hours, but I was determined to try. My first move was to go aboard the *Joseph Duhamel* and raise steam. I was glad to find that my *Maron* armed guard were there and would be making the trip with me. Nineteen of the original crew, led by André Lecoeur, remained, having volunteered for the Free French. All the others had chosen to be repatriated to France. Contraband Control had been trying to complete the crew by drafting Free French personnel from other French prizes. The extras consisted of two so-called deck officers, an engineer officer, an electrical officer, a stoker and four seamen, bringing our total complement to twenty-eight excluding the armed guard and myself. We could have done without the extra seamen, but urgently needed three stokers, three trimmers, at least one more engineer officer and three greasers. None was forthcoming.

Eugène Kerrault, our engineer, André and myself sat down together to work out a list of our requirements for three weeks. The engine room's needs were relatively clear-cut: water, coal, lube oils and greases, paraffin, packings and laggings, and rags and cotton waste. André reported that the thirty-four members of the old crew who had opted to go to France had taken most of the ship's bedding and mess traps (cooking utensils, crockery and cutlery) with them. The armed guard and I were short of warm clothing as we had had to leave the *Maron* with only what we stood up in.

I hurried to the Grand Hotel, to find a letter awaiting me from one of the senior Contraband Control officers with the name and address of the ship's agent and chandler. I laughed when I read, 'You sail tomorrow and so there is no time to lose.' I rushed to the chandler's to consult them over the list we had prepared. Then on to Naval Control, who gave me instructions to attend the convoy conference next morning at Naval HQ, accompanied by my first mate and chief engineer. Thanks to Contraband Control, within a very short time I managed to get hold of all the requisite charts, sailing directions, tide tables and, most important, the 1941 Nautical Almanac, without which it would have been impossible to navigate.

With a mad sort of fire brigade urgency I scurried from one naval establishment to another, never taking no for an answer. By the end of the day I had arranged for practically all our needs. When I returned to the *Joseph Duhamel* in the evening I was delighted with all that the crew and guard had done to make the ship ready for sea. I slept the sleep of the just that night.

We continued with our preparations at first light and by 9am had made fast at the coaling berth. It was difficult to accommodate the large tonnage of coal that we had been instructed to load, which was far greater than the ship's normal bunkerage. It was soon time for the convoy conference. I found it strange to be seated among professional merchant marine masters, chief engineers and mates, and to be treated as one of them. I was handed my set of Confidential Books and an envelope containing my secret orders, which I was told was on no account to be opened until we were at sea.

When we returned to our ship there were only forty minutes to go before we were due to sail and none of the supplies had been delivered; but within a few minutes the chandler's boat was alongside with everything we had ordered. We were in the middle of unloading all this stuff when our mess traps, bedding and warm clothing arrived.

Just then there was a hail from the mole. An officer from Contraband Control sprang on board and handed me a signal marked 'Urgent'. It read: 'The *Joseph Duhamel* is not, repeat not, to sail with convoy but is to return to Waterport anchorage forthwith. This cancels all previous orders.' I suddenly felt very, very flat. Even Contraband Control didn't know the reason for this last-minute change of plan but assured me that I was to retain command of the *Joseph Duhamel*. I was to live on board and continue my preparations, making ready to sail in the next convoy towards the end of the month.

Now that I had more time to reflect I realised what a blessing it was that our sailing had been so unexpectedly delayed. The *Joseph Duhamel* was not really in a fit state for a long voyage in the North Atlantic in the worst part of the winter. Our manning situation alone would have made such a trip hazardous. Over the next three weeks we were able to acquire one more engineer, a stoker and a trimmer from other French ships. I was able to persuade an RN engineer officer from the dockyard to make an inspection of our engine and boiler room. He wasn't very impressed with either, but as a result of his report quite a lot of valuable work was put in hand, including the replacement of a defective steam pipe.

The *Maron* arrived in Gibraltar on the night of 14 January, a good deal earlier than expected. I reported to the captain, whose manner was far from cordial. He was in a hurry to get ashore and told me to come and see him again the following morning. The long interview I had with him then was not a pleasant one, though I somehow managed to control my feelings and discreetly made no mention of the difficulties of my trip into Gibraltar without recognition signals or an Aldis lamp. I knew that Contraband Control had made a most complimentary signal to him about me because they had shown me the copy. But the fact that I was now designated merchant master in command of the *Joseph Duhamel* was like a red rag to a bull. It took all my powers of persuasion to get the captain's official blessing on my new appointment. However, he insisted on the return of my armed guard to the *Maron* and refused to recommend my promotion to full lieutenant on the grounds that he didn't know enough about me. RNVR sub-lieutenants over the age of thirty were automatically promoted after three months' satisfactory service and a formal recommendation from their commanding officer. An extra ring on my sleeve would have strengthened my authority as captain of the *Joseph Duhamel*, but the worst effect was the loss of several months' seniority.

My loyal but inexperienced armed guard party was replaced with a detachment of regulars from a destroyer, so I was much better off than before. Meanwhile the *Joseph Duhamel* had been legally condemned in the Prize Court and reregistered. Her new ship's papers showed her as a British vessel, owned by His Majesty King George VI (represented by the Ministry of Shipping), registered port, Gibraltar, and first master, Captain G M Hobday. We all had to sign proper merchant marine articles, including the French, who were to be paid normal British seafaring rates.

We were now to sail in convoy on the afternoon of 29 January and were told to allow for a voyage of twenty-three days. We had painted out the French colours on the sides and also the old port of registration, Fécamp, substituting Gibraltar. The new armed guard elected not to mess with the French crew, so we converted the radio operator's cabin on the bridge to be their quarters. I had indented for some timber to make the extra bunks and a rifle rack, work which was still in hand when we sailed. This arrangement improved our security should any trouble arise with the French, and it ensured privacy for both the British and the French sections of our crew. By the time we finished our final bunkering and provisioning the ship was heavily laden. Our freeboard at deck level amidships was a scant 15 inches, but I reckoned that it wouldn't take long to improve this, with the consumption of seventeen tons of coal a day, plus water and other stores.

Ours was a 'slow' convoy of some forty ships, escorted by an old China station river gunboat and a corvette. We headed a long way out into the Atlantic to avoid being spotted by enemy aircraft, before turning north. During daylight hours the convoy steered an evasive course in the form of a huge arc made up of zig-zags. I began to understand the reason for our seemingly excessive bunkering and provisioning. The normal distance from Gibraltar to the United Kingdom is 1,200 nautical miles. We ended up steaming just over three times that distance.

We had a generally uneventful voyage, suffering no enemy attacks, but we did have a few headaches. As I had feared, we couldn't consistently maintain a sufficient head of steam to keep up with the convoy. We lost it twice and were lucky to regain it. We had to use seamen as stokers and trimmers, and they were completely unskilled in this work. On many occasions I was forced to rouse the proper stokers from their off-watch sleep to help in the stokehold. I decided to disregard the evasive course routine and to guess what the straight mean course would be each day, although this meant risking an encounter with a U-boat. Each morning I would see the rest of the convoy steam away out of sight while we ploughed on alone following our straight course. As the afternooon wore on, I would scan the horizon anxiously in the hope of sighting the convoy before dusk.

On the last day or so before we approached the coast the

Aboard the *Joseph Duhamel*, Belfast Lough, February 1941.
Left to right: Eugène Kerrault, the author, André le Coeur, French crewman.

weather closed in. We became separated from the convoy before making our landfall, but managed a difficult noon sight. By midnight we were nearing the entrance to the North Channel, which separates Scotland from Northern Ireland. Because of the uncertainty of our position and the bad visibility, we were feeling our way with great caution, continually checking the log and taking echo soundings. It was six o'clock in the morning, pitch black and murky, when I got one of the frights of my life. A great light suddenly blazed overhead and for a split second I thought it was a star-shell. Then I realised it could only be Instrahull lighthouse, which is situated on a rock-bound islet in the middle of the approaches to the North Channel. We were very lucky not to have piled up on the rocks. During the war most lighthouses in the British Isles were not lit, but the crew in this one must have heard or seen us and unscreened the light. This near miss gave us a perfect navigational fix which enabled us to run on dead reckoning with confidence. There was still plenty of daylight remaining when the fog suddenly and obligingly cleared, and there, almost straight ahead, was Mew Island light, marking our journey's end – Belfast Lough.

We spent a week in Belfast, mainly at anchor in an open roadstead. An examination vessel had met us on arrival and a few fresh provisions were sent out to us next day, but neither the navy nor the Ministry of Shipping would recognise us as being their responsibility. However, the navy dealt efficiently with the ship, which they degaussed as a defence against magnetic mines. They also mounted two Lewis guns on deck for anti-aircraft purposes. I couldn't help thinking how remarkable it was that we had come all the way from Gibraltar to Belfast completely unarmed.

It was three days before anyone was allowed ashore – and then only me, accompanied by my leading seaman. It was a real treat to be on dry land again and to be able to stretch my legs after having been confined in such a small space for so long. As I tramped the streets of Belfast I recalled having read somewhere that a merchant master could attach a writ to his ship if the owners failed to meet their legal responsibilities in providing the necessary funds for port dues, crew's wages and provisioning of the ship. This seemed to be roughly the position I was in. A check in the reference section at the Belfast Central Public Library confirmed that a British ship's master had quite extraordinary powers in circumstances similar to those in which I was placed.

Greatly encouraged, I presented myself at the offices of a prominent firm of shipping agents and explained my predicament. They were only too eager to act for me and treated me with respect and courtesy. They would place a lien on the ship and cargo, obtain provisions and other supplies, and advance money. A chauffeur-driven hire car was quickly provided, which was available to me for the rest of my stay in Belfast. All these arrangements were subject to the agents visiting the *Joseph Duhamel* and inspecting her papers. They agreed to do this immediately. On the way I called in to report what was happening to both Naval Control and the Ministry of Shipping. They found it hard to believe that I was taking this drastic action, and it really shook them up. We soon had the ship provisioned and received our sailing orders. We were to proceed to Fleetwood and hand the ship over there. As soon as I knew this, I telephoned Helen and asked her to meet me on 23 February and to book us both into the North Euston Hotel. We had another day and a half in Belfast so I arranged some shore leave for my armed guard.

Visibility was bad again when we sailed for Fleetwood. Because of minefields, we had been routed north-about the Isle of Man, which meant skirting the rather unpredictable tides and currents of the Solway Firth. We had to wait for the tide off Fleetwood, and the pilot told me that we were probably the biggest ship that had ever gone into the docks there. It was 2pm before we berthed. As soon as I had dealt with the formalities required by a phalanx of officials, I placed an armed sentry on the gangway with strict instructions that no one was to go ashore and then went to see the resident naval officer. The RNO had been advised of our impending arrival, but had had no instructions regarding the disposal of the ship, cargo or crew. The old familiar story was repeating itself. It was clearly not the fault of the navy, but of my owner, His Majesty King George VI, represented by the Ministry of Shipping. The whole pot-mess was still further complicated by our having a French crew who had not yet been screened and approved by the de Gaulle authorities in London. Fortunately all the Fleetwood officials were very obliging and anxious to assist me. I told the RNO that so far as I could see, having signed the articles as master, I could not be released from the ship until the crew had been paid off. He agreed that I should appoint an agent, as I had done in Belfast, and that I could take any responsible action I saw fit as long as I kept him fully informed.

Helen arrived on the afternoon train from London looking as pretty as a picture. It was wonderful to see her again. She was very worried about how thin I had become – I had lost over two stone in weight and my well-cut uniform jacket now hung from me like a scarecrow. That evening when we went to the hotel bar for a pre-dinner drink the laughter and chatter stopped abruptly and someone said, 'That's the fellow.' A giant of a man came over and, putting his huge arm across my shoulders, told me in broken Teutonic English that he wanted to buy me a drink. He asked me how much I would be paid for bringing the *Joseph Duhamel* from Gibraltar, and took a lot of convincing that I would only be entitled to one shilling extra on top of my naval pay for doing that small job. He was an Icelandic trawler skipper and loaded with money. He was quite shocked that I should be treated in this way and solemnly peeled off a five pound note from a big roll and presented it to me. I was embarrassed, but his gesture was kindly meant and he wouldn't take no for an answer. The note represented nearly two weeks of my naval pay.

The Boston Deep Sea Fishing and Ice Company had been appointed as our agent and their Fleetwood manager, Basil Parkes,[1] was on board at nine o'clock on Monday morning to inspect our ship and cargo. He was favourably impressed and told me that there was practically no limit to the funds he was prepared to advance against such security. By a strange coincidence, we had met before – in the waiting room at the Admiralty Selection Board the previous year. He affixed a writ to our mainmast proclaiming that he had a lien on the ship and cargo.

I was in the RNO's office an hour later and got on the phone to London. I eventually managed to track down the minor Pooh-Bah in the Ministry of Shipping who was sitting on the Prize Court documents, which stated that the *Joseph Duhamel* was the Ministry's direct responsibility. The de Gaulle authorities didn't appear to be interested in our problems, and were certainly not prepared to send anyone to Fleetwood to screen our French crew. In due course we arranged the hire of a large motor coach to take them to

[1]Later Sir Basil Parkes (b. 1907). Chairman, Boston Deep Sea Fisheries Ltd, 1962–1980. President, British Trawlers Federation, 1966–1969.

London, accompanied by a posse of police and immigration officials.

While I was wrestling with these ponderous official affairs our industrious ship's agent was dealing with many others on our behalf. He had appointed a ship keeper, who would live on board henceforward. He had also prepared the necessary money and papers so that the ship could be paid off. I thought it best that the armed guard should be drafted to Portsmouth Barracks and that I should be sent back to *King Alfred*. The RNO approved, and said that he would have the necessary draft chits ready by morning.

I still recall my last day in Fleetwood with a certain nostalgia. That afternoon, in the presence of the proper officials, I solemnly paid off the crew. I prevailed upon the Ministry of Shipping representative to draw up fresh articles appointing himself as master so that I could be free to depart at any time, and handed over the ship's papers, together with all invoices for provisions, stores and services.

One of my last duties was to thank all those who had worked with me to contrive so much in so short a time. I made a final call on the RNO, for whom I had formed a high regard. He had been thoughtful enough to prepare and sign two certificates; one entitled the armed guard and myself to hard-lying money for the voyage from Gibraltar to Fleetwood, and the other awarded me command money from 7 January to 24 February. For an officer, hard-lying money amounted to three shillings a day and the command money was the same. I was now a man of comparative wealth thanks to these unexpected benefits, plus the Icelander's five pounds, to say nothing of the shilling I received for acting as merchant master.

That evening I said my farewells to the Frenchmen. It had not been an easy relationship for any of us, but common danger and hardship had brought us together. Some of them were quite emotional and embraced me.

On 26 February Helen and I travelled to London, sharing a compartment on the train with my leading seaman and the three able seamen. They had served me loyally and well, never questioning my orders, however extraordinary they must have seemed at times. It was largely through their fraternising with the French crew that any vestiges of doubt and suspicion had been removed. They were a credit to the navy, and I could now tell them so. We had a most pleasant trip together in a relaxed and friendly atmosphere. We went our separate ways on our arrival in London and parted with mutual good wishes.

I now realised that I must think hard about my future. There were two things I was quite sure about. First, nothing would induce me to rejoin the *Maron* and secondly I would do my damnedest, regardless of being over age, to join Coastal Forces.

3

After our journey down from Fleetwood, we went to stay with Helen's parents. Their new home couldn't have been more conveniently located. It was at Brighton, only two or three miles from *King Alfred*.

I had not been in touch with Commander Head, as I didn't want to risk being diverted elsewhere. He was the man I most wanted to see. I trusted him and he knew me reasonably well. He was surprised when I turned up, and concerned about my gaunt appearance. He told me to have a few days' rest and to let him have an informal written report covering my experiences from the time I left *King Alfred*. Helen's elder brother, Cameron Dahl, was manager of the Brighton branch of the Equity and Law Life Assurance Society, so I was able to dictate my report and have it typed at his office. I was very guarded in what I said, not only because of the demands of wartime secrecy but also because of a fear of appearing in any way vainglorious. These considerations resulted in a somewhat expurgated version, making no mention of some of the more colourful incidents. When I handed my report to Commander Head three days later, he asked me to sit down and wait while he read it. He did this with careful concentration and it seemed a long while before he laid it to one side of his desk. Then he looked up and said, 'Well, Hobday, this is a remarkable story which does credit to *King Alfred*. You have done well, and I am only sorry that I must direct you to report back to the *Maron* when she comes into Greenock off her next patrol. You have no alternative and I have no powers to arrange otherwise.'

I was not unprepared for such a comment and replied without hesitation, 'I'm very sorry too, sir, but I must refuse to return to the *Maron* under any circumstances and whatever the consequences.' I then added quietly, 'As soon as I leave your office, I intend to telephone the New Zealand High Commissioner, tell him the facts and place myself under his protection.' Commander Head reflected for a moment before replying, 'It would be wrong for me to say much but I think you would be wise to take that action because I must now report the whole affair to the Admiralty. Also, in my opinion, you need more time to recuperate before you can be fit for duty, so I'm granting you a further three days.' He then shook hands and wished me a good leave.

Immediately afterwards I telephoned the High Commissioner, explained the situation and sought his protection as a New Zealand national. I was in his office in London the same afternoon. Mr Jordan was very understanding, so there was no difficulty in securing his wholehearted support. The measure of his interest was reflected in a letter written to me by the New Zealand Associated Press correspondent, Alan Mitchell [reproduced overleaf]. Because of the reluctance I had to be publicised in any way during the war I politely refused his request.

When I reported back to Commander Head on 5 March he was all smiles. 'I've got good news for you, Lieutenant Hobday. You had better make time this morning to slip over to the tailors and get the second ring put on your uniform sleeves and another one on your epaulets. I've been in touch with the Admiralty, and they have agreed

Telephone: Correspondence - Central 7040. Advertising - Central 2686.
Telegrams: "Assoprez, London."

Alan W. Mitchell,
~~Miss A. E. Evans~~ Correspondent.

New Zealand Associated Press

23/28, Fleet Street
London E.C.4.

London Office of
New Zealand Herald / Weekly News — Evening Post — The Press — Otago Daily Times

March 13, 1941.

Dear Mr. Hobday,

When I visited H.M.S. King Alfred last night with the High Commissioner he told me about your experiences in the Mediterranean with the French trawler which you took as a prize.

I also had a chat with Commander J. S. Head, and he told me that he had a private report from you. Naturally, he was unable to show this to me as you had asked him not to, but he told me to write to you to ask if you would be good enough to give me a story for the New Zealand paper

I know these facts, for instance: (a) you were at H.M.S. King Alfred for five weeks; (b) that you boarded the French trawler with an armed party of five, (c) that you sailed to Gibraltar with its cargo of fish valued at some £10,000; (d) that you then brought it on to England although you had no master's ticket and only had five weeks' experience at H.M.S. King Alfred as a background. I also understand that you paid off the crew on arrival in England.

These are very sketchy facts and I should like a full graphic story You understand, of course, that anything I write has to be censored and if you like, I will send a copy of anything I propose to use first to you and to Commander Head. I will not quote you in any way so that it will not appear as if the story comes direct from you.

The High Commissioner and Commander Head are very anxious that the story should be told to New Zealand, and I think you will agree with me that your parents, and relatives and friends in the Dominion will get a great deal of pleasure from reading about your experiences.

Hoping that you will do this for me and will look me up when you are next in London,

With kind regards,
Yours very truly,

A.W. Mitchell

Sub-Lieut. G. M. Hobday,
H.M.S. St. Christopher,
c/o G.P.O.
LONDON.

Letter to the author from AW Mitchell of the New Zealand Associated Press, 13 March 1941.

to your promotion on my recommendation as from today. You have an appointment with the Second Sea Lord's department for three o'clock this afternoon. You are not rejoining the *Maron*. Subject to a satisfactory interview the Admiralty seem disposed to find a new appointment to your taste.' My faith in Commander Head had not been misplaced, and I guessed that he had given me the additional leave largely to allow himself time to pull a few strings on my behalf.

My interview in London went well. Their Lordships were obviously pleased with me, and after some discussion agreed to waive the age limit so that I could be appointed to Coastal Forces. They telephoned the liaison officer for Coastal Force appointments, instructed him accordingly and directed me to his department. Here I found a very angry RNVR lieutenant-commander, who was a petty bureaucrat of the most miserable type. As he prepared my draft chit, railway warrant and other documents he kept muttering to himself such remarks as, 'They've no right to go over my head like this,' and 'It's wrong to waive the age limit – I've never heard of such a thing.' Everything was done with very bad grace. As I went out of the door he fired his last shot: 'I'll see you never get a command!'

I couldn't get back to Brighton quickly enough to pass on all the news about my promotion and posting to Helen. The thought of another inevitable parting didn't seem so bad now, because I would be based somewhere in England. I had one more free day before departing for HMS *St Christopher*, the Coastal Force training base at Fort William in Scotland. We made the most of those precious hours together, but I did find time to see Commander Head to thank him for all his help and to let him know about my success in securing my Coastal Force appointment.

I arrived at Fort William on 8 March and was billeted in a hotel which had been taken over by the navy. The two-week course, which started the next morning, was a good mixture of practical and paper work and was both vigorous and intensive. Great emphasis was placed on physical fitness; each day we turned to at 5.30am and, rain, hail or snow, were on the parade ground in shorts, singlet and sandshoes at 6am for PT, after which we had a three-mile run, cold shower and breakfast. It was very cold while I was at Fort William. Nearly all my classmates were many years younger than myself, and I could see that I must make some dispositions to ensure that I survived the course. Within a day or two I had found a bakery shop and tea shop, kept by a nice old biddy, situated in a side alley about a hundred yards up the main road from where we started our run. Most mornings I managed to break off and disappear up this alley, to sit in the comfortable warmth of the bakery while my gallant comrades continued with their strenuous training. When I heard the thundering herd returning down the hill I would quietly rejoin them, and of course always came in among the leaders. The sprint with which I used to finish earned me the nickname 'Flash' Hobday, and this stayed with me in certain quarters for some years.

I made many friends on the course and served with, or met, most of them at some time during the three and a half years I spent in Coastal Forces. Coastal Forces was a special, tight-knit 'private' navy and we had much in common with the other well-known private navy – the submarine service. In both cases officers and men were selected to a high standard and had to volunteer for this type of duty. This was quite unlike the situation in general service where personnel were drafted to any type of ship as necessity demanded. There was no room for any supernumeraries like cooks or stewards in our ships: every man-jack had to have specialist skills. We were treated in the same way as submarine crews and allowed submarine 'comforts', namely extra energy rations, such as sugar and cocoa. We wore the same rig at sea and were issued with submarine 'frocks', which were heavy white rollneck sweaters that reached down to the knees. The crews also enjoyed the submariner's perk of a rum issue, which was served neat to all ranks, in contrast to general service where it was watered down two to one, except for senior ratings. These privileges, plus our more interesting and exciting way of life, sometimes generated a degree of envy in general service ships, which gave rise to such comments as, 'Coastal Forces – damned Costly Farces!'

A tremendous spirit of enthusiasm activated everyone at Fort William. The training flotilla there was made up of all types of Coastal Force craft – MLs (motor launches), MTBs (motor torpedo boats) and MGBs (motor gunboats). Our instruction alternated between ship-handling exercises at sea, and classroom studies and lectures. One of the major classroom subjects was aircraft and ship recognition, on which we had almost daily sessions with models, photographs and silhouettes. Instant identification of friend or foe and knowledge of the armament and capabilities of the different types of ships and aircraft were an absolute essential when operating in Coastal Force craft. The many types of mines used by the enemy and ourselves were also studied with great care.

Before I completed my *St Christopher* course I was notified that I was to be appointed as first lieutenant of ML 110, 1st ML Flotilla, based on Great Yarmouth in Norfolk. I was able to telephone Helen and make arrangements to meet her there on 24 March 1941. We both had difficult journeys in crowded wartime trains, with several changes. My journey lasted thirty-six hours. Helen was sitting wearily in the lounge of the Victoria Hotel. We made straight for our bedroom, but we were no sooner there than the air raid sirens began their banshee wailings. Soon afterwards there was a knock on the door; a porter had come to show us to the hotel air raid shelter in the basement. I was too tired to leave my comfortable bed, and Helen stayed on with me. It wasn't long before the bombs started to fall and we heard the drone of aircraft and the barking of cannon fire. From the window Helen saw low-flying German fighter-bombers shooting up buildings on the promenade. There were fires everywhere. The hotel next door received a direct hit and Wellington Pier immediately opposite us was blown in half. This was the first major raid on Great Yarmouth, where there were some sixty days and nights of these hit and run affairs.

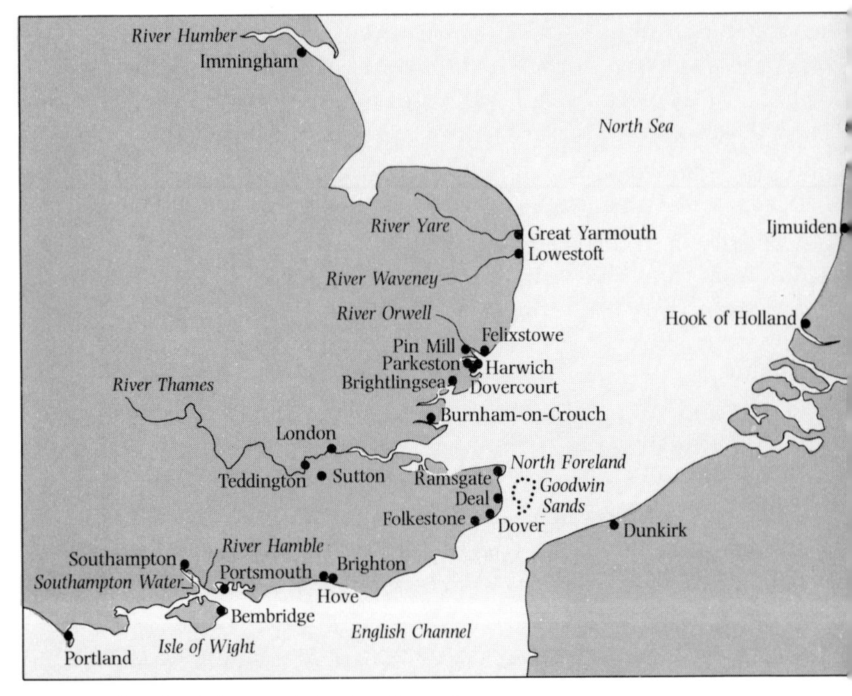

The North Sea and the English Channel

In the morning, after our somewhat unsettled night, I advised Helen to leave, as Great Yarmouth didn't seem to be a very healthy spot. She thought otherwise, and when I set out for the 1st ML Flotilla base she made her way to the main Air Raid Precautions centre, which was in a state of confusion, full of shocked mums and screaming children whose houses had come down about their ears or had caught fire. The air raid wardens were doing their best but were rushed off their feet. Helen, being a trained teacher, sized up the situation at once and quickly went off from shop to shop, buying with her own funds such things as Plasticine, crayons and painting books. When she got back to the centre, she gathered the children round her at a big table and soon had them occupied and reasonably happy. She continued this labour of love for some days but I finally had to insist that she returned to the south, not only for her own safety but also because her presence in Great

Yarmouth was a worry to me and disturbing my own work and duties.

When I arrived at the Fish Quay, where the 1st ML Flotilla was based, I went aboard 110 to meet the captain and crew. My captain was also newly appointed and the previous captain and first lieutenant spent the next two days handing over to us. 110 was an 'A' type ML, the first I had seen. Only a few of these were built before they were succeeded by the new 'B' type Fairmile ML. Like all Coastal Force flotillas, the 1st ML comprised eight boats in two divisions of four. The 1st division had 'A' boats and the 2nd division 'B's. I thought the 'A's were splendid vessels. They were 110 feet long, very seaworthy, and quite the fastest of all ML types, powered by three twelve-cylinder Hall-Scott aero-engines to give a top speed of 30 knots. They were equipped with Asdic and depth charges for anti-submarine work, and armed with a number of twin-mounted Lewis guns plus a 3-pounder mounted forward. Like the 'B's, they were maids of all work and used for convoy ecort, anti-E-boat patrols and air-sea rescue as well as minelaying and minesweeping.

The new captain of ML 110 was a 'permanent' RNVR lieutenant,[1] well trained and versed in naval matters and nautical jargon, and always impeccably turned out. He was a nice enough chap, but completely lacking in any aptitude for manoeuvring a ship. This serious handicap was made the more dangerous by a speech hesitation when he was under stress. Furthermore, when he got himself into one of his usual mix-ups, his brain seemed to freeze and he would then, rightly or wrongly, invariably give the order 'starboard wheel'; inevitably this became his nick-name. We were soon the most feared ship in Great Yar-mouth. The crew and I acquired an almost ballet-like agil-ity in leaping about with fenders whenever ML 110 left the quay or returned to her berth. We were lucky to have an

[1]Author's note. Most RNVR officers were appointed for the duration of hostilities only. 'Permanent' RNVR were the equivalent of the Territorial Army and before the war spent some time each year training in Royal Navy ships and shore bases.

 ML 110: steering position.

‘A’ class ML in the North Sea, 1941.

excellent coxswain, who often averted catastrophe by disregarding his orders and taking things into his own hands.

The port of Great Yarmouth lies on the River Yare, which is narrow and swift flowing. On the ebb the outgoing tide can reach six knots. It is no place for an incompetent ship-handler. On one memorable occasion when we were leaving our berth, the captain, instead of holding on to the forespring to let the current cant out the stern and then going astern into mid-stream, gave the order, 'Let go everything!' Finding that the tide kept the bows pinned against the quay, poor old 'Starboard Wheel' increased power in a desperate effort to get the ship away, but with no success. We continued to scrape along the quay at ever-increasing speed. Fortunately there was nothing berthed ahead of us except a small fishing boat. The two men on board were watching us with some interest, which soon changed to a look of incredulous horror. Just before we hit them they managed to vault the six or seven feet from their deck to the top of the quay. With our razor-sharp, hardened steel bow, we sliced through their boat from stern to stem, the two halves falling apart like segments of an orange. It was a dreadful moment and we all felt guilty and ashamed. Needless to say, Starboard Wheel didn't retain his sea-going command for very long and was soon transferred to other duties.

In the succeeding weeks we spent most of our time as a convoy escort. The East Coast convoy system was the largest in the world, averaging sixty to eighty ships per day, northbound and southbound, with London as the focal terminus. Because of the German occupation of Holland, Belgium and Northern France and also because of the neutrality of Eire, nearly all shipping had to be routed round the north of Scotland and up and down the East Coast. These convoys were our most vital lifeline.

From London to a little north of the River Humber, the North Sea is shallow and full of dangerous shoals, which made escort work difficult and allowed no flexibility in the courses that were followed. This section of the convoy

ML 110: twin Lewis guns and depth charges.

route became known as 'E-boat Alley'. The main E-boat base was only ninety miles away at Ijmuiden in Holland. Because of the constrictions of the shoals and the narrow deep-water channel, the convoys could be formed into only two columns, which stretched for about seven miles, presenting a tempting target for the E-boats. These would make massed attacks from the safe seaward side of the shoals and, after letting go ten to sixteen torpedoes, turn tail and head for home. The area was an ideal place for ground mines. The Germans laid them nearly every night by E-boat and aircraft.

Our normal routine was to steam out of Great Yarmouth in the late afternoon, test our guns, probably receive the odd air attack and pick up the northbound convoy about twenty miles off the coast. We would escort it throughout the night and break off at daylight to enter the Humber and berth at Immingham, which is about ten miles

56 Some of the crew of ML 110.

up-river from Grimsby. Once there, we would refuel, clean the guns, carry out self-maintenance routines, make any repairs and rest up until the next day. We would then steam out of the Humber, pick up the southbound convoy, stay with it through the night and, when daylight came, make for our base at Great Yarmouth. This escort work in E-boat Alley was a most demanding and tiring job. All one's senses had to be kept on a strained alert for many hours on end. There were few convoys where we didn't lose some ships by E-boat attacks or mines, and we were in great demand as rescue craft.

Coastal Forces were being expanded very rapidly, so that first lieutenants like myself were continually being promoted to take command of new ships as they were completed. After many officers who were junior to me and less competent had been promoted to command their own ships, I began to feel pretty bloody-minded and recalled the parting remark of the liaison officer at the Admiralty, 'I'll see you never get a command!' Then, just as I had got to the end of my tether and might have done something very stupid, I had a stroke of luck.

I had come off convoy duty that morning feeling very tired and fed-up. When I left the ship in the early evening for my shore leave, instead of making for the Queen's Hotel, which was the usual rendezvous for Coastal Force officers, I decided to go to the Star Hotel where the minesweeper officers used to congregate. I knew I would be bad company and just didn't want to see anyone. I sat in the lounge sipping a drink near a large group who were joking together and spinning yarns. After a while, one of the younger officers started to tell a story that held everyone's interest. It was an account, told in some detail, of the capture of the *Joseph Duhamel*, the trip to Gibraltar and the voyage to Belfast and Fleetwood. I could hardly believe my ears and went straight over and asked the young officer where and when he had heard that story. 'That's quite simple,' he said. 'It's a standard passing-out pep talk given at *King Alfred* by Admiral Kekewich.'

ML 106.

Next morning I saw the base commander privately. I told him how I had come to hear the *Duhamel* story the previous night and mentioned the Coastal Force liaison officer's comment. He seemed surprised to hear this. He said that I had been recommended for command a number of times and that he could not understand why I had been passed over. I asked if I could have two days' leave to see Admiral Kekewich to let him know what had happened to the 'enterprising' officer whose virtues he had extolled. The commander was not having this, but said he would go up to London and see Admiral Kekewich himself. Two days later he sent for me and gave me the good news that I had been appointed to immediate command, and would receive my three shillings a day command money forthwith. There was no new ship available for some time, but meanwhile I was to act as captain of the flotilla leader's ship, ML 106. I was very pleased about this. I got along well with the Senior Officer of the 51st ML Flotilla, Lieutenant-Commander Lochner, who was a man of exceptional ability. Young as he was, he had become a director of the well-known firm of electrical engineers and manufacturers, Crompton-Parkinson.

Within a week or two of my joining 106, three of the 'A' boats (including 106 as leader, but not, thank God, 110) were detached from the 1st ML Flotilla and converted to carry out a very special and important job. Our assignment, which was top secret, was to lay a new type of mine close inshore on selected parts of the Dutch and Belgian coasts, and in particular to block the various mouths of the Rhine. We moved down to HMS *Beehive* at Felixstowe, which was to be our operational base. Someone had shown a good deal of imagination in naming all the Coastal Force shore bases after insects. Great Yarmouth was HMS *Midge*, Portsmouth was HMS *Hornet*, Dover, HMS *Wasp* and so on.

At Felixstowe our depth charges were removed and special racks were installed for the new mines, which were acoustic and magnetic ground types. These were fitted with an anti-sweeping device, which could be set so that it would detonate only after minesweepers had passed over it as many times as the number set in the mine. They were filled with a new explosive, more powerful than TNT, and were cylindrical in shape, weighing about a ton apiece. The mines contained a soluble plug of compressed salt, which made them inactive until the salt dissolved by immersion for fifteen to twenty minutes. Each 'A' boat carried six of them.

The two most important elements of good minelaying were extreme accuracy in the positioning of the minefield and in the spacing between the individual mines. Exact positioning was essential not only to ensure the greatest hazard to the enemy but also to ensure the safety of one's own ships. Eventually all minefields have to be cleared and it is vital then to know exactly where they are. Each type of mine had its own 'critical distance'. If laid any closer together one detonated mine would destroy the whole field. Conversely, if the mines were laid too far apart, it would be possible for a ship to pass through unscathed.

East Coast convoy.

With our 'A' type ML minelaying group we took infinite pains to achieve the greatest navigational accuracy. We ran our ships over a measured mile at various engine revolutions and calibrated the speeds. This was done many times to take into account changes in our draught with different mine loads and varying amounts of fuel. We also swung each ship to correct for compass error before every trip, with mines on and mines off. Under Lochner's leadership, every preparation had to be thorough and every contingency allowed for. We knew there would be little hope of our survival if we were caught by the enemy on this sort of job. It was Lochner who conceived the lovely stratagem of arranging for an announcement on the BBC morning news after each operation, 'Last night our aircraft laid mines in enemy waters.' It must have been largely due to this that we were able to work without being detected, but naturally we didn't like others being given the credit.

The Coastal Force base at Felixstowe. Note the damage to the flour mill, caused when a Hampden bomber crashed into it in 1940.

Our first lay was three-quarters of a mile off the Hook of Holland entrance to Rotterdam. The visibility that night was too good for our liking, but we had an unexpected piece of luck. When we were about five miles off our objective a hundred or more German aircraft, returning from a raid on Britain, passed low overhead and fired their recognition signal. Now we not only knew the current German recognition signal, but also the aircraft had in effect identified us to the shore as 'friendly'. We proceeded to carry out our lay with confidence.

The smallest sounds carry very clearly over the water on a still night and under such conditions we had to be stealthily quiet when close inshore. We ran very slowly so as not to create much wash, using only one specially silenced engine. Orders were given in sign language or in a sort of stage whisper. Sometimes it was all so conspiratorial, particularly when certain extraneous gestures were added, that it was difficult not to get a fit of the giggles. We often wore slippers or crept round the decks in stockinged feet. We always had a feeling that we could be seen, and waited with a strained expectancy for searchlights to beam on to us and the shore batteries to open fire. I'm quite certain that we were sometimes spotted, but the Germans must have assumed that we were some of their own coastal patrol units. One night we were so close in, near Neiuwpoort in Belgium, that we could hear drunken laughter as several soldiers came out of an inn and also the slamming of their car doors and the whirring of the self-starter. The coxswain whispered to me, 'I could do with a pint myself, sir.'

My last minelaying trip ended in a peculiar manner. We were nearly home, but as we turned to shape our course for Harwich we actuated a mine, probably an acoustic one. These were always more likely to be triggered when a ship was changing course or speed. Fortunately we must have been some distance off because we didn't suffer any great damage. The stern lifted as it received the mighty thump and we lost a few bits of planking from the outer skin of the hull. The rudders and propellers were unharmed but we sprang a number of leaks, which the combination of our powered bilge pump and our hand pumps couldn't quite keep pace with. The whole of the after compartment was badly strained and there was no one place that we could plug effectively. We needed a slipway to carry out repairs, but when we signalled Harwich they replied, 'Regret no slip available here. Proceed Felixstowe for orders.' When we berthed at *Beehive* there was a power pump ready on the quay and the welcome information that slipping and repair facilities were being prepared for us at Lowestoft. As soon as he was assured that the power pump was gaining on our leaks, Lochner handed the ship over to me. He was required in London for a conference at the Admiralty later that day.

I decided to lighten the ship before sailing, and off-loaded as much ammunition and stores as I dared. The weather was deteriorating and when we cast off it was already

Lieutenant-Commander R A Lochner (left) and the author on the bridge of ML 110.

blowing Force 6, which increased to Force 7 before we reached Lowestoft. We found we made least water by cruising at only ten knots, so it was already getting dark as we steamed up the River Waveney to Brooke's boatyard. Brooke Marine was an exceptionally large yard and built fishing trawlers and MLs as well as yachts. As we came alongside we were met by a party headed by Pip Brooke himself. They had everything ready for us – except the vital slipway. This could not be cleared until the morning but we were thankful for the two portable motor water pumps which had been scrounged from the fire brigade. We found that one of these had all the capacity we needed – we took the precaution of holding the other on standby – and we soon had a dry ship again. Having completed all arrangements with the yard and made quite sure that the ship was in no danger, I handed over to the first lieutenant and the navigator, both competent and experienced officers, and went ashore to stretch my legs and relax.

I returned to ML 106 a couple of hours later to find a very irate commander who was in charge of the Lowestoft Coastal Force base, together with some members of his staff. We had words together and he threatened to have me court-martialled for deserting my ship while 'in a sinking condition'. I refused to be browbeaten by him in front of the ship's company. It didn't improve matters when I pointed out that the charge he had made was patently absurd. Next morning, as soon as 106 was safely on the slipway, I managed to get through to Lochner, who had just returned to *Beehive*, and reported my fracas with the commander. Lochner just said tersely, 'Leave it with me. I'll see that that so-and-so is dealt with.' He was not a man to mess about and he went straight to the top, Rear-Admiral Goolden at Harwich, who was responsible for Lowestoft and other ports and naval establishments nearby. It was not long before Admiral, Harwich, was on the telephone to Commander Coastal Forces, Lowestoft, saying that he considered any grounds for my court-martial were quite baseless and that he wouldn't countenance such proceedings. Fortunately, within the week, I received instructions from the Admiralty that I had been appointed to command ML 339, then completing at Burnham-on-Crouch, and was to report forthwith to Captain Coastal Forces, Brightlingsea, Essex.

4

As I made my way to Brightlingsea, my happiness about taking command of a brand new ship of my own was tinged with some apprehension. I had heard some fearsome stories about Captain Coastal Forces, who was alleged to run a very Spartan course and rule everyone with the proverbial rod of iron. It was true that Captain Farquharson was a perfectionist and a strict disciplinarian but he was a man that most of us came to like and respect. He was a saturnine Celt with bushy tufts of black hair on his cheeks. Because of this he was often referred to (out of earshot) as 'Bog Face'.

The yacht-building firm of Aldous at Brightlingsea were now constructing Coastal Force craft. Their yard facilities, jetties and moorings were invaluable for our familiarisation and working up course on 'B' type MLs, which included a good deal of practical ship handling and manoeuvring. Each newly appointed commanding officer was joined by his first lieutenant and key members of his crew. I was well satisfied with my team. Vibert, the first lieutenant, came from the Channel Islands. Short and spare in build, he always looked you in the eye and possessed great self-confidence. Petty Officer Godfrey was a fine figure of a man, standing over six feet. He was a good all-round sportsman, had a likeable and easy manner, and was skilled in knots, splicing and fancy rope-work. Bird, the chief motor mechanic, was young and eager, and turned out to be a genius with internal combustion engines.

Most of the rest of the men who joined later were novices, but they soon settled down and within a few weeks we had a happy and efficient ship, manned by a crew that I was proud to command. ML 339 came out top in every subsequent admiral's inspection and her operational record became the best in our flotilla.

When we were entertaining on board or were engaged on a monotonous task, such as painting the ship, we often

Members of the crew of the newly built ML 339, 1941.
Sub-Lieutenant ACH Vibert is on the extreme left.

used to ask our senior gunnery rating, Leading Seaman Lee, to play for us. Lee was a veritable virtuoso on the piano accordion and could hold people spellbound with his magic. When a big inter-service talent competition was organised we persuaded Lee to enter and were delighted when he progressed through the local, county and regional contests, eventually to win the final in London early in 1942. C B Cochran, the great impresario, thought so highly of him that, when he presented the prizes, he publicly offered him a contract as soon as he could be released after the war. Sadly, he did not live long enough to take up this offer.

There were many advantages in serving in Coastal Forces, not the least being the reasonable possibility of achieving command of an independent sea-going ship in a comparatively junior rank and the chance to experience in full the authority and responsibilities that only the com-

The author on the bridge of ML 339 at her builders' yard, Burnham-on-Crouch. The guns have not yet been mounted.

mand of a naval vessel can give – together with all the accompanying privileges and status. Furthermore, because Coastal Force craft usually worked in twos and threes rather than in complete flotillas or half-flotillas, ordinary commanding officers were often called upon to lead units on operations. As captain of one of these ships, you were generally responsible for making your own decisions and you were briefed directly by higher authority, instead of having to depend on rumour or hearsay for information. You had the responsibility, too, for the general well-being of your crew, the granting of special leave, and recommendations for promotions and decorations.

In October 1941 we left Brightlingsea to join ML 339 at her builders' yard near Burnham-on-Crouch. It was normal naval practice to send the captain elect to stand by his new ship in the final completion stages; during this time he could not only acquire a thorough knowledge of her constructional details but could influence the placing and design of some of the smaller items of equipment. Once the ship had been fitted out the captain went to sea for contractors' trials. On return he rendered a defects list to the builders, who had a similar list of their own. When it was considered that all faults had been rectified, the captain and his naval crew accompanied the builders on sea-going acceptance trials and if everything was in order he signed the acceptance papers for the ship on behalf of the Admiralty. Until that moment the ship was the responsibility of the yard.

The 'B' type MLs were designed by the Fairmile Marine Company in conjunction with the Admiralty and were the first Coastal Force craft to be mass produced by means of one of the earliest kit systems. With these prefabricated sets the building of the 'B' type MLs was not confined to boatyards, and many were turned out by furniture manufacturers and the like. Although the 'B's were apparently identical, there was a good deal of variation in the soundness of construction and the quality of workmanship and finish, depending on where they were assembled. They were beautiful little ships with good built-in accommodation for officers and men, who lived permanently on board. They were slower than the 'A' boats, being powered by twin 650 hp Hall-Scott petrol engines, which had a top speed of 19–20 knots, but they could stay at sea a lot longer than the 'A's, thanks to the well-designed hull, which required much less power to drive it at the lower cruising speeds. They were excellent sea boats although very lively in bad weather. Like the 'A's, they were equipped with Asdic and depth charges, but carried more armament than their predecessors.

ML 339 was built by Wallasea Bay Yacht Station, one of the better yacht yards, with pride and craftmanship in spite of the kit system employed. We were very pleased with her and she proved herself to be a remarkable ship. After successful acceptance trials it was the custom for the builders to provide a sort of banquet on board for the officers and crew by bringing hampers stuffed with provisions as well as wine and champagne. The brothers Zabell, who owned the yard, came down from London with a few very ordinary sandwiches and no wine. They did, however, produce one bottle of whisky. After pouring out two meagre drinks for us, they resealed the bottle and departed with it. We were astounded at their meaness. We heard later that a similar incident had occurred after the acceptance trials of the previous 'B' type ML from this yard. After the same sort of sandwich lunch, the two owners had departed and the ML sailed for Brightlingsea. The commanding officer found a telegram waiting for him there. It read, 'Please return the partly consumed bottle of whisky which was inadvertently left aboard your ship.' The officer concerned happened to be a good Scot who liked his Scotch. Without hesitation he telegraphed an immediate reply, 'Regret that whisky inadvertently left on board has inadvertently been consumed.'

As soon as we had completed our working up period at Brightlingsea, we sailed for Pin Mill on the River Orwell, to be degaussed for protection against magnetic mines. It was a tranquil scene when we dropped anchor there, and the war seemed far away. The famous old hostelry, The Butt and Oyster, lies close to the water's edge, adjoining a mole where Thames barges often berthed while awaiting a favourable tide. Vibert and I had secretly arranged for our wives to meet us at The Butt and Oyster on our arrival. After showing them our new ship we returned to the pub for dinner. As 339 could not be left for long without an officer on board, Vibert and I had decided to take alternate watches. I was to take the first one and he was to relieve me at 1am. So when he and his newly-wed wife retired soon after dinner, Helen went on board with me to snatch some sleep in one of the wardroom bunks until it was our turn to relax ashore.

At 11pm I heard a lot of shouting on deck. I rushed up to find that our seaboat, with a number of men returning from shore leave, was rapidly disappearing into the murk on the fast flowing ebb. The crew had apparently lifted oars to pass them on board, but nobody had made fast and they were now headed for Harwich and the open sea. It was our only boat and no ship movements were allowed on the river after dark. I hailed the crew of a Royal Engineers' motor fishing vessel moored nearby. They only had a small rubber dinghy but they got me ashore and helped me liberate a decent sized rowing boat from a boathouse. We set off down river to search for our missing boat and liberty men. Luckily we found them only a couple of miles away, aground at a shallow point on a river bend, and we were soon safely back aboard 339. I could do no less than invite the Engineer major and his lieutenant, who had been so helpful, below for a drink, and only hoped that Helen would stay snuggled down in the bunk in the shadows. However, as we entered the wardroom she sat bolt upright in her frilly nightie, and there was no mistaking the fact that the bunk's occupant was a shapely young woman. There was a short silence and then without batting an eyelid the major said to Helen, 'And how long have you been at sea, my boy?'

After their departure I decided to take the risk of keeping Helen on board for what remained of the night – I would probably have been court-martialled if I had been found out. We were thus able to leave the young lovers undisturbed and in due course I rowed Helen ashore in a Turneresque dawn.

Two days later, degaussing completed and compass readjusted, we steamed into Great Yarmouth to berth at the 1st ML Flotilla base, receiving a great welcome from our sister ships. Lieutenant Kenneth Cutler, previously half-leader, was the new Senior Officer of the flotilla. Blond and blue-eyed, he had been a first mate in the crack Royal Mail Line before the war and I knew him to be a competent officer. To my pleasure, I found that our new half-leader was Lieutenant Basil Bourne. Basil was a delightful man with a cultured and wealthy background. He had been educated at Wellington College and St Catharine's College, Cambridge, and had been a partner in a firm of stockbrokers in the City. A keen athlete who kept himself very fit, he was conscientious and decisive in command but off duty was sometimes inclined to be a little hesitant in his speech.

Lieutenant BL Bourne.

We had become good friends at the Fort William training base.

The new flotilla engineer officer was an interesting character. Engineer Lieutenant the Earl of Craven was a sports car enthusiast and just before the war had married one of the daughters of Mrs Meyrick, London's night club queen. As a result he had been forced to resign his commission in the Coldstream Guards. He then volunteered for the navy, feeling confident that they would want to utilise his skills with high-speed petrol engines. The navy accepted him all right but appointed him to a coal-fired tug as a stoker, probably with the idea of taking some of the starch out of him. After some months, they trained him as an engineer officer, giving him special courses on Hall-Scott and Rolls-Royce Merlin aero-engines, which were being widely used in Coastal Forces. Of average height and slimly built, he always stood very erect. He had delicate features, slightly ginger hair and a milky complexion that would have done credit to a schoolgirl.

By now the flotilla comprised eight 'B' type MLs. Our duties, which were still mainly concerned with convoy escort, included rescue work, identifying and dealing with floating mines and undertaking anti-E-boat patrols. In the few months I had been away from E-boat Alley much had been happening. So many ships had been sunk by E-boats, mines and aircraft that the main channel and approaches were becoming littered with wrecks, which were a depressing sight and a grave menace to navigation. They were being cleared as far as possible with high explosives, and the rescue tugs accompanying each convoy had developed a routine of towing sinking ships out of the channel and dumping them alongside or even on top of previous wrecks. A confidential Admiralty order had been issued to the effect that merchant crews were not to be treated as survivors until a naval officer from one of the escorts had satisfied himself that there was no possibility of getting the damaged ship to port even after emergency repairs. It had been found, in some cases, that crews abandoned their ships immediately they had been hit and there is no doubt that many valuable cargoes were saved as a result of this order. This is no reflection on the proven courage of British merchant seamen, but they did lack the training and organisation of the Royal Navy.

Their inherent coolness in an emergency and their lack of discipline were both brought home to me in an irritating manner one night when a rusty old Geordie collier was mined in convoy and started to founder rapidly. I had 339 alongside in a matter of minutes, by which time the collier's main deck was nearly awash. 'Jump for it!' I shouted. No one moved. 'Are you bound for the Tyne?' they asked. 'No, the Humber.' 'Aw, bugger that!' Fortunately for the collier one of the rescue tugs, which happened to be on its way to the Tyne, arrived on the scene to tow it clear of the channel and take the crew on board.

Although air raids on Great Yarmouth were less frequent than they had been, the Luftwaffe continued to make quite a number and also occasionally attacked us in daylight when we were sailing to rendezvous with northbound

'B' class MLs in line ahead, North Sea.

convoys. These air attacks never seemed funny at the time but some of them provided us with a good laugh afterwards. We had just left harbour late one afternoon when a twin-engined Messerschmitt 110 swooped in low on our starboard beam and raked us with cannon shell. Then, turning sharply, he attacked from our port side, this time dropping two bombs as well as letting us have another dose of his cannon fire. He made his final run so low that his remaining bomb hit the water at a very shallow angle and ricocheted off the surface to pass close over our heads

ML 339: cleaning the forward 3-pounder gun.

as we stood on the bridge. As the coxswain remarked afterwards, 'You could have hooked it down with a walking-stick.' We watched, fascinated, as it wobbled on its erratic path but were mightily relieved when it plunged into the sea without exploding some fifty yards to port.

In the autumn of 1941 all New Zealanders serving in the Royal Navy were invited to transfer to the newly formed Royal New Zealand Navy, which had previously been known as the New Zealand Division of the Royal Navy. This sounded very attractive as we were not required to leave the Royal Navy ship in which we were serving and our income would be practically doubled. The New Zealand rates of pay were appreciably higher than the British and no income tax was levied. Strangely enough this inducement attracted little interest and nearly all of us refused to transfer from the Royal Navy. We were, after all, volunteers and would have felt ashamed to be paid more serving in the same ship and exposed to the same dangers as our British brother officers.

Considerable rivalry had developed between the Coastal Force bases at Great Yarmouth and Lowestoft. At Great Yarmouth, in addition to our ML flotilla, we now had a flotilla of the new 'C' type motor gunboats, while at Lowestoft they had an ML flotilla and a motor torpedo boat flotilla. Sometimes a combined unit, made up of boats from both ports, was needed. There was much chagrin at Great Yarmouth one evening when, owing to last-minute defects, we couldn't contribute the requisite number of boats for a joint operation. ML 339 was out of service, requiring repairs to her hydraulic steering gear. However, acting on impulse, I rigged emergency tiller steering and put to sea with our unit. Fortunately for our patrol we didn't encounter the enemy that night because it would have been difficult to fight an action with that cumbersome equipment. I little knew then that my tiller steering effort was to have an important sequel some three years later.

In late November 1941 four MLs were selected for a special mission and detached from the 1st Flotilla. Our flotilla leader, Ken Cutler, commanded the unit for this expedition, so it seemed certain that we were going to be involved in something important. We were ordered to Parkeston Quay, the peacetime passenger and cargo terminal at Harwich, where we found a complete flotilla of eight modern fleet minesweepers, plus two similar Dutch navy ships, berthed alongside. These fleet sweepers looked like small frigates and were quite unlike our familiar friends, the sweepers converted from fishing trawlers.

We were briefed as soon as we arrived and learnt that the operation was a hush-hush one and was likely to take several weeks. The fleet sweepers were to clear minefields close to the Dutch coast in daylight. Our MLs were to accompany them on every trip. Our main tasks were to identify and subsequently destroy certain mines as they were swept, to screen the sweepers against E-boat attack and to act as rescue craft if any sweepers were mined or bombed. For the first time in our experience the RAF was participating, by providing two Spitfires to give continuous fighter support during daylight hours. The endurance of the Spits was short and they could only cover us for twenty or thirty minutes before being relieved. We were surprised to learn what a large number of aircraft were needed to sustain this operation but were glad to have them with us.

We worked long and hard under the most appalling discomfort. To spend the maximum hours of daylight in the sweep areas, we had to sail at about 3am and didn't get back until nine or ten o'clock at night. The winter hardened earlier than usual and was one of the most severe ever recorded on the East Coast. It was sheer torture to be steaming at 16–17 knots through the penetrating cold, with salt spray flying over the bows and freezing solid on our faces. Those terrible conditions tested our hardiness and endurance to the utmost, and we used all sorts of dodges in our efforts to stay warm. The oven was kept filled with large potatoes heating in their jackets; these were doled out, two at a time, to all hands, to be placed piping hot in the pockets of their duffle coats. Soup and cocoa were handed round frequently. I recall that when I was on the bridge I dropped a mug of near-boiling cocoa as it was being passed to me; the spilt contents froze into a chocolate-coloured wafer as they hit the deck. Except for the first lieutenant and myself, all hands were able to take short turns below to thaw out in front of the roaring coal fire in the big cooking stove in the galley. For once we envied those of our shipmates whose duties kept them in enclosed spaces below decks – the cook in his galley and the engine room staff all cosy and warm with their beloved motors.

While in the sweeping areas, we had to maintain the same course and speed as the relatively big ships we were escorting and were unable to make any variations to ease the motion of our boats in heavy weather. If we were stuck with a nasty beam sea, the well-known propensity of 'B' type MLs to roll their guts out was given full play and further added to our wretchedness. Twice we actually rolled our deck-mounted depth charges under, which brought back disagreeable memories of the *Maron*.

Towards the end of December the cold grew so intense that our ships became frozen in at Parkeston Quay and couldn't be moved without risk of damage to their thin wooden hulls. Harwich had always been an ice-free port and this was the first time in living memory that the water

The author photographed on board ML 339 after a long night on escort duty.

had been frozen over. Unexpectedly we were able to celebrate Christmas in harbour. Instead of lurching around at sea in dangerous waters, with all our senses heavily strained and alerted, we were immobile, safe and relaxed, and almost as free of duties as if we were on leave. Vibert and I had lost no time in arranging for our wives to join us on Christmas Eve, for what we had reason to hope would be a stay of several days. We were able to book the girls into the excellent seaside hotel at Dovercourt, a resort on the North Sea coast near Harwich.

After the hazards and discomforts we had endured we were more than ready to let off steam on Christmas Day. Things really started moving from the time I reversed roles with the steward, as is the traditional Christmas Day custom in the navy. He rang the bell and ordered drinks in the wardroom, which I served on a silver tray while he burlesqued being in command. He roundly ticked me off for taking my time and for not polishing the glasses. This change of roles inspired Helen to swop clothes with Vibert. Ken Cutler came aboard and, after a few drinks in the wardroom, we all went down to the messdeck to have a Christmas toast with the crew. Some wag there suggested to Helen that she should get rid of Ken's much-prized beard. Without more ado she ordered our gallant flotilla leader to be seized. He was forced into a chair and, while one of the men held a mug of tepid water for her, Helen lathered his face and neck thoroughly and scraped off his Drake-like beard and moustache amid scenes of great enthusiasm. This must have been a painful operation for Ken but he bore it with apparent good grace, while I thought to myself, 'Goodbye to any hope of promotion in this flotilla.'

Parties continued throughout the day. At a particularly riotous one on board the senior Dutch minesweeper, the *Jan van Gelder*, her captain, Baron van Geen, was debagged. In the evening those who were still on their feet gathered in the lounge of the Dovercourt hotel. Harwich was a garrison town and this hotel was also frequented by the army. In spite of the uncomplimentary terms which the navy applied to the military, such as 'brown jobs', 'pongos' and 'the brutal and licentious soldiery', and the banter which flew when we encountered each other, relations between the two services were always good.

When our party was well under way some genius produced a china pot plant holder, into which we poured our various drinks to make a loving cup. While this evil concoction was being handed around, the commandant of the garrison, Lieutenant-Colonel Sir John Carew Pole,[1] CO of the 5th Battalion, Duke of Cornwall's Light Infantry, passed through the lounge. Full of bonhomie, we invited him to drink from the loving cup, a privilege which he was ill-advised enough to decline. Quite spontaneously he was pushed to the floor and rolled up in the carpet. Six of us bore the bundle upstairs. It then occurred to us that it

[1] Lord-Lieutenant of Cornwall, 1962–1977.

might be a good idea to get away from the hotel with all possible speed and we beat a hasty retreat to our ships.

About ten o'clock next morning the flag lieutenant boarded us to advise me to make ready to receive Admiral Goolden and the garrison commandant on board at eleven. When we met them at the gangway, the ship was spotless and we were all impeccably turned out. As soon as our distinguished visitors had been piped over the side the admiral addressed me. 'What I have to say to you had better be said in private.' With a distinct feeling of trepidation I led the way to the wardroom. They would not sit down and the admiral came straight to the point. 'I understand, Hobday, that you were concerned in the disgraceful affair last night when Sir John was assaulted and wrapped in a carpet. I have come with Sir John this morning so that you can make your apologies to him in my presence.' The incident no longer seemed so funny and my apologies were most gladly and sincerely made. The admiral continued, 'I have discussed this matter with the commandant and he agrees with my view that an apology is not enough. You must be punished as well and the punishment must fit the crime. As it was not convenient for Sir John to join you in a drink last night, you can supply us with a glass of your best Scotch.' They both burst into laughter. When I saw them over the side half an hour later I thought how effectively they had dealt with me.

We continued our minesweeping in severe weather for several weeks after Christmas. On one of the last days of the operation, the fleet sweepers located a line of moored mines and made a preliminary sweep, using cutters, to free a few so that we could close and identify them. Once this had been done the sweepers knew exactly how to proceed. They removed the cutters and used normal paravane sweeping wires, which rolled the mines over so that they became self-detonating. Our fleet sweepers then advanced in a single echelon formation and swept the line of mines with such accuracy that they didn't miss one of them in the whole seven-mile length of the field. It was remarkable to watch the mines explode, one after another, each in turn hurling a great fountain of water as much as two or three hundred feet into the air. One of our Spitfire boys must have had a hell of a fright when he swooped down low just as we started the sweep and was almost enshrouded in spray.

We never discovered the purpose of all this minesweeping, but, interesting though it was at times, we were glad when it was finished and we were able to return to Great Yarmouth. Here we resumed our work on convoy escort and anti-E-boat patrols. Basil Bourne was promoted to command a 'C' type motor gunboat, a move which automatically made me flotilla half-leader. Basil's Number One, Bob Harrop, succeeded him in command of his motor launch.

Not long afterwards four of us were once again sent away on detached service. This time we were based at Ramsgate on the Channel coast. The seafront there presented a most depressing sight. The windows of all the houses and hotels were boarded up or sandbagged, and the beaches were covered with barbed wire and rows of anti-tank traps. We carried out night patrols that ranged from Folkestone to the North Foreland at the entrance to the Thames. During the several weeks we spent on this assignment we had no encounters with the enemy but there was one incident that is worth recording. When we were on patrol in thick weather near Deal I suddenly heard breakers and realised that I must be close to the dreaded Goodwin Sands, which are as hard as iron and have been a graveyard for ships from time immemorial. I sheered off promptly, but as I did so several massive wire ropes scraped along our hull and bridge with an ominous twanging. It was so unexpected and eerie that it really put the wind up me, as only some menace from the unknown can. My immediate reaction was to ring down 'Stop' on the engine room telegraphs, but 339 carried her way for some time. A great mast loomed out of the night only a foot or two to port. My hair nearly stood on end as it dawned on me that we were passing between the mast and supporting shrouds of a wreck, and must have had very little clearance between our keel and deck obstructions such as cargo winches. We were extremely lucky not to have had the bottom torn out of our ship.

At Ramsgate we were separated from our base maintenance and repair facilities, and the condition of our ships soon suffered accordingly. We sweated hard to keep 339 at a high level of efficiency. She was always ready for sea and when other ships became unserviceable we had to substitute for them, with the result that we carried out many patrols over and above our stint. I have no fond memories of Ramsgate and indeed the whole ship's company was delighted when we were ordered back to Great Yarmouth, which we had come to regard as our second home. We were in high good humour when we headed north in daylight up the East Coast. About ten miles south of Harwich, as we were turning to port to avoid an obstruction caused by two sunken ships in the swept channel, we actuated an acoustic mine. It exploded like a thunderclap, lifting our stern out of the water and scattering a few small pieces of planking from our thin outer skin. Everything happened so quickly. The screws revved up and the ship shook violently as we bounced back to our normal displacement. By this time we were approaching the other wreck, which was lying directly ahead of us. Our engines were running smoothly and there was no reason to think that we had suffered any real damage. I shouted down the voice pipe, 'Starboard 20!' Nothing happened, so I shouted 'Emergency, hard-a-starboard!' Still nothing happened. I realised that the ship would not answer her helm, and I instinctively gave the order 'Stop starboard, half astern starboard!' This had the desired effect and we just managed to avoid the wreck.

I sent for Dicky Bird, my chief mechanic, and together we examined the steering gear with the coxswain. We could find nothing wrong with it, but there was no load on it when the wheel was turned and we came to the conclusion that we had lost both rudders when we had set off the mine. I signalled Harwich for permission to enter and put in a request for divers. It's a tricky job trying to keep a ship on course by engine revs only and an even trickier one to manoeuvre alongside a quay into a tight gap among crowded shipping, but we managed better than we expected. The divers confirmed the loss of our rudders and we were towed to a vacant slip for repairs. A few days later we were back in Great Yarmouth, to resume the familiar routine of convoy escort duty and anti-E-boat patrols.

Soon after our return to Great Yarmouth, Vibert, whom I had strongly recommended for command, left to take over one of the 'B' type air-sea rescue MLs. By a strange coincidence his replacement, Derek Eastgate, a Royal New Zealand Naval Volunteer Reserve lieutenant, came from Takapuna, Auckland, only three miles away from my own birthplace. Eastgate proved to be extremely capable and became popular with our crew and with his brother officers. Because of slight balding on top he was nicknamed 'The Egg'.

On 11 June 1942 one of our MLs returning from operations signalled the sighting of a strange-looking floating mine. It turned out to be a new magnetic variety and was soon hauled gently ashore on a beach at Corton near Lowestoft by a mine investigation squad. Lieutenant-Commander Edwards and Ensign Howard, a young United States Navy officer who was on temporary duty as an observer, went to work to defuse it, following established routine. Using non-magnetic tools, they shouted their intentions before each step to other members of the squad who were sheltering behind a sandbag screen. Thus if a mishap occurred the same mistake would not be repeated next time. The risks were considerable because new types of mine were often booby-trapped. This mine exploded during the attempt to defuse it, blowing the two men to pieces. The remains were unidentifiable and were put into a common casket. As I had been friendly with both these fine officers it was decided that ML 339 should be used for their burial at sea. The United States Embassy in London sent some marines to make up a joint firing party with members of my crew. There were many wreaths, both official and private, which were floated over the casket as it disappeared into the depths. It was a moving ceremony and we were all very silent as we returned to harbour.

With the diversion of the greater part of the Luftwaffe to Russia and to a lesser extent to the Mediterranean, air raids on the East Coast had been tapering off during the past few months. On the other hand, E-boats were being built in ever increasing numbers. Their attacks became more frequent and more could be deployed each time.

On 6 October 1942, as I came into my berth on a crisp autumn morning after escorting a southbound convoy down the coast, a RNVR lieutenant leapt aboard and handed me an official envelope marked 'Urgent'. The message inside instructed me to hand over command of 339 temporarily to the bearer, and to 'report forthwith' to HMS *Seahawk* for a special course on anti-submarine warfare. Shortly afterwards, Bob Harrop came to tell me that he was going on the same course. *Seahawk* was actually a shore establishment at Ardrishaig, situated on Loch Fyne near the entrance to the Crinan Canal, fifty to sixty miles by boat from Greenock. We set off about noon, and I was glad to have such a cheery travelling companion as Bob. I turned around to take a last look at my ship. I hated leaving her, if only for a few weeks. She was the apple of my eye, and with her splendid crew was literally my pride and joy.

5

We travelled throughout the night, and our train connected with the steamer at Greenock about ten in the morning. After a pleasant cruise up Loch Fyne we reported to HMS *Seahawk* where we were each issued with a bicycle and directed to our billets, which were in a decent private hotel on the lochside. We thought the bicycles a bit of a joke at first, but on the flat ground round the straggling township of Ardrishaig they proved to be the most convenient means of transport.

Early the following morning I had an unexpected telephone call from Ken Cutler. He was brief. 'Bad news, Geoff. Your ship was sunk last night – torpedoed. Eastgate and Lee killed – all the rest OK. Sorry.' He rang off. I stood there, utterly shocked, with the receiver still in my hand. I'd lost two fine men, friends, and my lovely ship of which I had been so proud. It was the only time she had been to sea without me and she was the first Coastal Force craft to be torpedoed. I learned later that she had been shepherding a northbound convoy when E-boats had made a massed attack. They fired a dozen or more torpedoes simultaneously at the convoy and ML 339 had been unlucky enough to stop one. Our course started that day, but I saw the captain of *Seahawk* and sought leave to go back to Great Yarmouth to look after my men. He was understanding but quite firm in his refusal. He assured me that they would be well cared for at the base and I later received a letter from the coxswain which confirmed this.

The discipline at *Seahawk* was rigid and everything had to be done at the double. I was made captain of our class, which comprised some twenty to thirty officers. In turn I appointed Bob Harrop as my Number One. Fair-haired and with a florid complexion, Bob was a big man both in build and in character. Before the war he had managed his family's large furniture store in the heart of Manchester and was proud of being a North-countryman. He did nothing by halves, and if he gave of his friendship, as he had done to me, it was absolute. He soon imposed his own strict discipline on the class, and I suspect, by the deference and respect I was accorded, that he had told them a lot of favourable lies about me.

The course was a good one, with the right mixture of classroom lectures and practical work operating Asdic in boats on the loch – with an occasional trip in a submarine thrown in for good measure. It was most refreshing to live ashore again after the cramped quarters of an ML.

Two weeks after my arrival at *Seahawk*, I was appointed to command one of the new 'D' type motor gunboats – the largest ever built for Coastal Forces and with formidable firepower. I was instructed to report to the Senior Officer of the 19th MGB Flotilla at Teddington near London. My classmates threw a terrific farewell party that evening and congratulated me on getting 'the finest command in Coastal Forces'. Next day, in defiance of the authorities, Bob Harrop and the entire class assembled on the quay to see me off.

I stayed at Helen's flat at Carshalton overnight and telephoned my new flotilla leader first thing in the morning. It was none other than Lieutenant 'Mickey' Thorpe, whom I had come to know well when he was Senior Officer of the 16th MGB Flotilla operating from Great Yarmouth. His flotilla, to which Basil Bourne had been posted, had their base at the same quay as the 1st ML Flotilla about a hundred yards downstream of us. Mickey was a man I admired and respected. Less than two months before, to rescue survivors from MGB 335 as she lay on fire, he had engaged twelve E-boats single-handed and then fought a long rearguard action with them in daylight as he retired towards Great Yarmouth at the best speed he could make. It was rumoured afterwards that he was recommended for the VC for this act of sustained courage, and many thought he fully deserved the ultimate award. Instead, he was given an immediate DSO, a decoration rarely conferred on a junior officer and signifying heroism little short of VC quality.

Lieutenant E M Thorpe on leave at Weymouth in March 1941.
With him is Gwyneth Acland, a family friend, and the Thorpes' kitten, Figaro.

It was a sound Coastal Force practice in those days to give Senior Officers of important new flotillas the privilege of choosing the commanding officers of each boat. I felt honoured to be selected by such a man as Mickey. When I reported to him on board his new 'D' class MGB I met a number of other COs, including Basil Bourne and Lieutenant 'Tufty' Forbes. Tufty had also served in the 16th MGB

Flotilla and we had often visited one another's ships. It was Tufty's MGB which had been set on fire in Mickey's action with the E-boats. Squat and powerfully built, Tufty tended to be bull-headed and irascible, yet he had a good sense of humour and there was something very likeable about him. He owned a chain of tobacconist's in London.

Mickey gave us a run-down on the new ships, which were also being built in a motor torpedo boat version. He had no idea where we might be based – it might even be overseas. Over the next week or two we familiarised ourselves with every detail of the 'D' class MGB. Seven more were in course of construction for our flotilla, but the most advanced were those allocated to Tufty (MGB 646), Basil (MGB 645) and myself (MGB 643), and it seemed likely that we would be the operational vanguard.

We were impressed with Mickey's prototype ship, MGB 644. She was 115 feet in length, very beamy and powered by four 1,250–1,500 horsepower supercharged Packard Merlin engines. In addition there were auxiliary engines to drive two 12½-kilowatt generators which supplied current for the all-electric galley. MGB 644's guns could pour out some thousands of rounds per minute of well-assorted fire, and the 2-pounder pom-pom forward, the twin .5 Vickers machine guns either side of the bridge, together with the twin-barrelled Oerlikon automatic cannon amidships, were all power-mounted with joystick controls. The heaviest gun was the after-mounted 6-pounder. Additional to the main armament were no fewer than eight Vickers 'K' .303 machine guns, mounted in pairs, while the armoury contained several short automatic rifles for boarding or landing. Although the ship was not equipped with Asdic, depth charges were carried. But the wonder of wonders to us in those early days was to find that she was fitted with radar.

Despite our enthusiasm, we foresaw that with so much jammed into so small a space we were going to have problems in maintaining such complex ships at a high standard of efficiency; manoeuvring in harbour would be difficult too, as all four screws rotated in the same direction. With three or four officers and twenty-five to thirty crew on board we would be living cheek by jowl. One important compensation for the commanding officer was that for the first time in Coastal Force craft he was provided with a small cabin, which incorporated a writing desk for dealing with the ever-increasing flow of bumf.

I was soon on my way to the Isle of Wight for the completion stage and acceptance trials of my new ship. I'd struck it lucky again because MGB 643 was being built by Woodnutt's, which had a reputation for fine workmanship. The Isle of Wight was a restricted area more or less reserved for permanent residents, so there was plenty of vacant holiday accommodation. I stayed in an excellent hotel near Woodnutt's yard at Bembridge. The officers of 41 Royal Marine Commando were billeted there while in training to go overseas. They were a lively and agreeable lot and we had few dull moments in our spare time. Like us they had no idea where they would be based. Curiously enough, although at that time there was no hint of an Allied invasion of Sicily, I apparently remarked to the CO, Lieutenant-Colonel Lumsden, as I said goodbye, 'I'll see you in Sicily then.' He reminded me of this when we met there soon after the landings the following year.

Unlike the builders of ML 339, Woodnutt's really did us proud after our Admiralty acceptance trials, with lashings of champagne and a great variety of delicacies. Our final trials were carried out on the Solent. As we finished we went into the fuelling berth to fill up to our maximum capacity of five thousand gallons of 100 octane petrol. During fuelling, no heating, cooking or smoking was allowed, and I was very grateful when the CO of a steam gunboat, moored just ahead of us, came along and invited me aboard for a cup of coffee. The bulkheads of his wardroom were covered with Peter Scott wildfowl paintings. When I remarked on them, he laughed and said, 'I find them easy to live with, but I'm probably biased because, you see, I *am* Peter Scott.'[1]

[1] Lieutenant P M Scott, then commanding SGB 9. Author of *The Battle of the Narrow Seas* (London, 1945). Later Sir Peter Scott. Artist and ornithologist.

'D' class MGB.

Although I was delighted with 643 as a ship, I was far from happy about some of the crew. They were generally a surly and slovenly lot, many of whom were obviously rejects from other ships or drafted from detention barracks ashore. By this stage of the war, with the continuous expansion of the navy, we were starting to scrape the bottom of the barrel to find crewmen of experience. However, I had a competent and likeable Number One in Lieutenant Norman MacLeod. 'Mac' was an Oxford classicist and the greatest master of the English language – written and spoken – that I have had the privilege to know. He was a spare sandy man who never seemed to get rattled no matter how tough the going was and had one of those accentless Scots voices which are a delight to the ear. His family ran one of the most prosperous woollen mills in Yorkshire. He had had a short time in the Fleet Air Arm as a fighter pilot before serving with Mickey Thorpe in 'C' type MGBs. The Fleet Air Arm had found him too expensive to retain as he had written off a couple of their fighters within a few weeks of going solo.

My Number Two, Sub-Lieutenant Claude Holway, a cheerful, chubby Londoner, was a pleasant fellow but lacked experience in Coastal Force craft. The brightest spot in our ship's company was the engine room staff, led by my ex-339 chief motor mechanic, Dicky Bird, whose ability was quite outstanding. Between us we had overcome many difficulties to get him transferred to 643 but he had been as keen to be with me again as I had been to have him.

Christmas 1942 was nearly upon us when we motored up the River Hamble, off Southampton Water, to take up moorings in what was to be our fitting-out and working-up base. The captain of the base had some important news for us. 'The 19th MGB Flotilla is urgently required in the Mediterranean,' he announced, 'and 643 is to be made ready as quickly as possible for shipment with three others by heavy lift ship to Gibraltar. You had better indent for issues of tropical kit. Because of the urgency, there can be no question of any of you having leave under any circumstances.'

When my disaffected crew, who were already resentful of the régime of hard work and discipline I had imposed on them, were told that there would be no Christmas leave and that we were going overseas, they became openly truculent and it seemed that serious trouble could erupt at any moment. The flash point came next morning when we were berthed at the end of Bursledon jetty to load stores and equipment. I took the opportunity of going ashore to deal with some official business. On my return, as I clonked my way over the transverse planks of the long wooden pier towards the ship, the quartermaster left his place at the gangway, turned away and busied himself with coiling a rope down neatly on the deck. He still had his back to me when I reached 643, although he had seen me from afar and had heard me approach. It was an act of studied insolence in front of other ratings working on deck. I shouted to one of them, 'Tell the first lieutenant and the coxswain to report to me immediately.' They were on deck instanter. The quartermaster remained on his knees by the rope, trying to brazen it out – now somewhat uneasily. I pointed to him. 'Place him under arrest and charge him with gross dereliction of duty and behaviour insulting to an officer. The witnesses are here and I'll hear the case within the hour.' A case of this type should have been dealt with by higher authority ashore, but I couldn't afford the time this would have entailed; so I contented myself with giving the quartermaster the maximum punishment allowed to a lieutenant in command – extra work, fourteen days' stoppage of shore leave and seven days' stoppage of pay.

Then the balloon went up. Next morning the coxswain handed me eleven written requests to leave the ship for transfer elsewhere. Excluding the coxswain and my loyal engine room staff, this amounted to about half of the crew. The malcontents were a cunning lot and they knew full well that if I forwarded these requests to higher authority, as I must, my competence and suitability to hold a command could be brought into question. The only alternative would be for me to have a friendly and placatory chat in the hope of persuading them to withdraw their requests. But such a course would give them the whip hand, and I was not prepared to compromise on this issue, whatever the outcome.

After some thought, I resolved to take the war right into their camp. I briefed my officers, sent for the coxswain and gave orders to clear the lower deck and for the entire ship's company to be fallen in, in the correct rig of the day, on the jetty alongside the ship. 'I want every man-jack on this parade, with the eleven request men in a separate rank in front,' I said. 'Number One, you are to send down a message to me when the parade is ready.' Soon I could hear them all assembling. When the messenger arrived I decided that a little suspense wouldn't do any harm, so I kept them waiting for a few minutes. After the first lieutenant had called the parade to attention, I addressed the eleven out front very briefly and deliberately. 'Your requests to leave this ship will be forwarded today, accompanied by my written comments, in which I will state that all your applications bear the same date, indicating an organised conspiracy tantamount to mutiny. I will also state that most of you have been trouble-makers ever since you joined this ship. As some of you may not realise what a serious position you are in, I am prepared to give you a chance by allowing any of you to withdraw his request within an hour from now.' Within the hour they had all withdrawn their requests. I had had the whole ship's company present, partly to make my little address seem more convincing, and partly to ensure that the whole crew would know how the matter had been dealt with instead of getting a garbled version from the request men. My handling of this affair seemed to have the desired effect. In a few days MGB 643 had become a happy ship and remained so thereafter. Our working-up programme further consolidated matters by generating a loyal team spirit that was good to see.

In the New Year we heard that Bobby Craven had been appointed as Engineer Officer to our 19th MGB Flotilla. I welcomed this as we had got on well together when he was Engineer Officer to the 1st ML Flotilla. Then we learnt that a serious problem had arisen over our passage to the Mediterranean. No heavy lift ship could be made available to us, so we would have to get there under our own steam. On the face of it, this wasn't possible as the minimum distance greatly exceeded our fuel range, but it had been calculated that by fitting additional tanks on deck to increase our fuel capacity by 3,000 gallons and by running slow ahead on only one engine at a time, we should be able to make Gibraltar in between five and seven days depending on the weather. There were obvious risks in this plan, but, as always in wartime, these had to be balanced against advantages gained in getting our ships on operational station without delay. No one really knew how those lightly built craft would stand up to the huge seas often encountered in the Bay of Biscay; and we couldn't take any but the shortest route, which meant that we would be exposed to attack from the German navy and aircraft from the Luftwaffe's French Atlantic bases for at least half our voyage. Added to these factors was our inability to defend ourselves, as the 100 octane deck tanks could be fired or exploded by flashes from our own guns.

In mid-February 1943 MGB 643 was ordered in to Thornycroft's yard at Southampton for final overhauls and fitting-out. We scrubbed bottom, and heavy duty brackets and wire rope bridles were fitted fore and aft so that we could tow or be towed in case of breakdowns on our ocean voyage. We also repainted the whole ship in the light blue-grey of the Mediterranean Fleet.

Next weekend Colonel Sheldon, managing director of my old company, brought Helen down to Southampton and treated us and my fellow officers to a slap-up dinner at Southampton's newest and most luxurious hotel, the Polygon. The colonel had lost an arm and won the DSO serving with the Royal Engineers in the First World War. It was a great evening and the following morning we invited him to visit us aboard 643 before he returned to Kent. He was amazed to find so much fighting capability packed into so small a ship and shocked to see how bombing had ravaged the ancient town of Southampton, with more than half of it reduced to rubble.

Early in March we were ready to set out for Milford Haven to rendezvous with MGBs 644, 645 and 646. I had been separated from the others ever since I had taken over 643 at Woodnutt's yard and I was looking forward to joining up with them. We were to sail in company with the *Brecon*, a new Hunt class destroyer built by Thornycroft's and routed for Liverpool down the Channel and up the Welsh coast. The 360-mile trip was a pleasure cruise in calm seas and sunny weather and we arrived at Milford Haven on 13 March in high spirits. I reported at once to Mickey Thorpe, to find him pessimistic about the chances of 643 sailing with his unit for Gibraltar. The other three 19th MGB Flotilla boats, together with a number of 'D' type MTBs, had been at Milford for some weeks and were all well advanced in their final preparations. The base staff were fully occupied with these ships and could spare little time to work on 643. In any case there was no room for us at the fitting-berth and we had to anchor out in the bay. As day succeeded day, with no berth and little help from the base staff, it seemed clear that it wouldn't be possible for us to make the passage with the others.

Mickey said, 'It's bad luck, Geoff, but you'll just have to wait here for the next lot to assemble and join up with us later in the Med. We're the guinea pigs, and if our trip is successful all the larger Coastal Force craft will be steamed out in future.' 'Look, Mickey,' I replied, 'I'm not going to give in to these circumstances, and come hell and high water I'm determined to sail with you.' Mickey smiled and said, 'I don't see how you can do it – but I wouldn't put anything past you.'

I attended all the conferences that Mickey held with the COS of the other ships and carefully observed the routines being carried out aboard them by the base staff. We then carried these out in 643 on our own. The base staff had now almost become fellow conspirators in my efforts to get 643 ready, and by the time we had a berth practically all that remained to be done was to fit the deck petrol tanks. Mickey sailed with his 'convoy' of three MGBS and one MTB (651) on 23 March, and I virtually stood over engineers, plumbers and fitters while the petrol tanks were installed and final general checks made. I signalled that I would be running trials. When these were satisfactorily completed I made a further signal that I intended to join my unit and, some twelve hours after his departure, started my pursuit of Mickey. I knew the courses his ships would be steering and that their speed would be limited to 9 knots. I estimated that by steaming 643 at 18 knots, I would overhaul them about 220 miles away in nine hours.

In the event it all seemed very straightforward and we did catch up with our colleagues exactly as anticipated, but over that distance it would have been easy enough to have missed them and I was greatly relieved when we sighted them on the horizon to port. The weather during our three days in the danger zone of the Bay of Biscay could not have been better for our safety – overcast with poor visibility. Because of this we were not spotted by any German aircraft and suffered no enemy attack. On the other hand a Force 7 wind blew for the first two days, creating quite a sea in addition to the big ocean swells. It was frightening at first to watch the effect on our lightweight hulls. From the bridge we could see the terrible flexing strains to which our boats were subjected, with the bows twisting one way and the after section the other. We were heavily loaded with additional stores, spares and ammunition, but our greatest anxiety was the effect on our stability of the extra 12 tons of weight on deck resulting from the 3,000 gallons of petrol plus the tanks. We used up the deck fuel first and gradually regained stability, but were very conscious of the fact that empty petrol tanks are a much greater explosive hazard than full ones.

Twice a day we had to notify Mickey of the exact amount of fuel remaining, and also advise him of any defects in hull, engines or equipment. By the time we entered the 'safe' zone off the Spanish and Portuguese coasts the weather started to clear and moderate, and we were soon enjoying sunshine, warmth and comparatively calm seas. Our fuel reserves were so favourable that two days out from Gibraltar Mickey stepped up speed to 12 knots and twelve hours later to 15 knots. It was then that MTB 651 fell out of formation and signalled that she had fuel feed problems. Mickey ordered me to tow her and reduced the speed of his convoy to 12 knots. Soon afterwards he signalled me again and asked whether I could manage 16 knots with my tow. We had no problem in doing this and we even experimented successfully with increasing the tow speed to some 18 knots, thus establishing some useful data in case of future need.

On 29 March we arrived off Europa Point and made our recognition signals so that we could enter Gibraltar Harbour. My 'tow' came through to me on his loudhailer to report that he had managed to carry out some repairs on his fuel system and could proceed under his own power. I replied that I would come alongside to pick up his end of my towing cable. Unfortunately, before I could do this, he let it go and the whole 200 feet of heavy wire rope hung down vertically, nipping hard on our transom. Not being equipped with winches or capstan or even roller fairleads, we had to rig blocks and tackles; it took nearly all hands over an hour to recover the cable and stow it on board. However we were in a jubilant mood about the success of our voyage from Milford and this annoyance was soon forgotten.

6

Seldom before in its long history had Gibraltar been of such vital importance to Britain and her allies as it now was. This great fortress and naval base had recently become an air base and staging post as well. The Gibraltar isthmus was covered with Allied aircraft of all types, including no fewer than fourteen fighter squadrons. Gibraltar had also been made the headquarters of the newly appointed Supreme Commander in the Mediterranean, General Eisenhower.

Our own operational base was to be Bone, where we were urgently required. We learnt that the situation in North Africa was far from good. The Eighth Army's rapid advance westwards after Montgomery's victory at El Alamein in October 1942 had petered out by the end of December. The First Army, advancing from the east after the Allied landings in Morocco and Algeria on 8 November, had also come to a halt. The Germans and their Italian allies held the whole of the North African coast between Bone and Tunis. They also controlled the European coast from France to Turkey as well as all the Mediterranean islands with the exception of Malta and Cyprus. It was impossible for our merchant ships to pass through the Mediterranean, so most of Montgomery's supplies still had to be sent via the Cape of Good Hope, a hazardous voyage of nearly twelve thousand miles.

For my part I felt very privileged to have command of a fighting ship in the Mediterranean at that historic time. The sole purpose of motor gunboats like MGB 643 was to seek out the enemy in his own waters. The only other types of ships that fulfilled this role were MTBs and submarines. I found the prospect before us both exciting and sobering. I knew we would be in the thick of it, frequently fighting actions, and that inevitably we must suffer casualties.

Facilities for Coastal Force craft at Gibraltar were very limited and we had no opportunity to rest after our tiring passage from England. We were forced to carry out self-maintenance on our ships for two or three days before sailing on 2 April for Algiers some five hundred miles away. We decided to retain our deck tanks so that we could make the voyage at high speed on all four engines. The weather was bad and we had a very rough ride to

Bone.

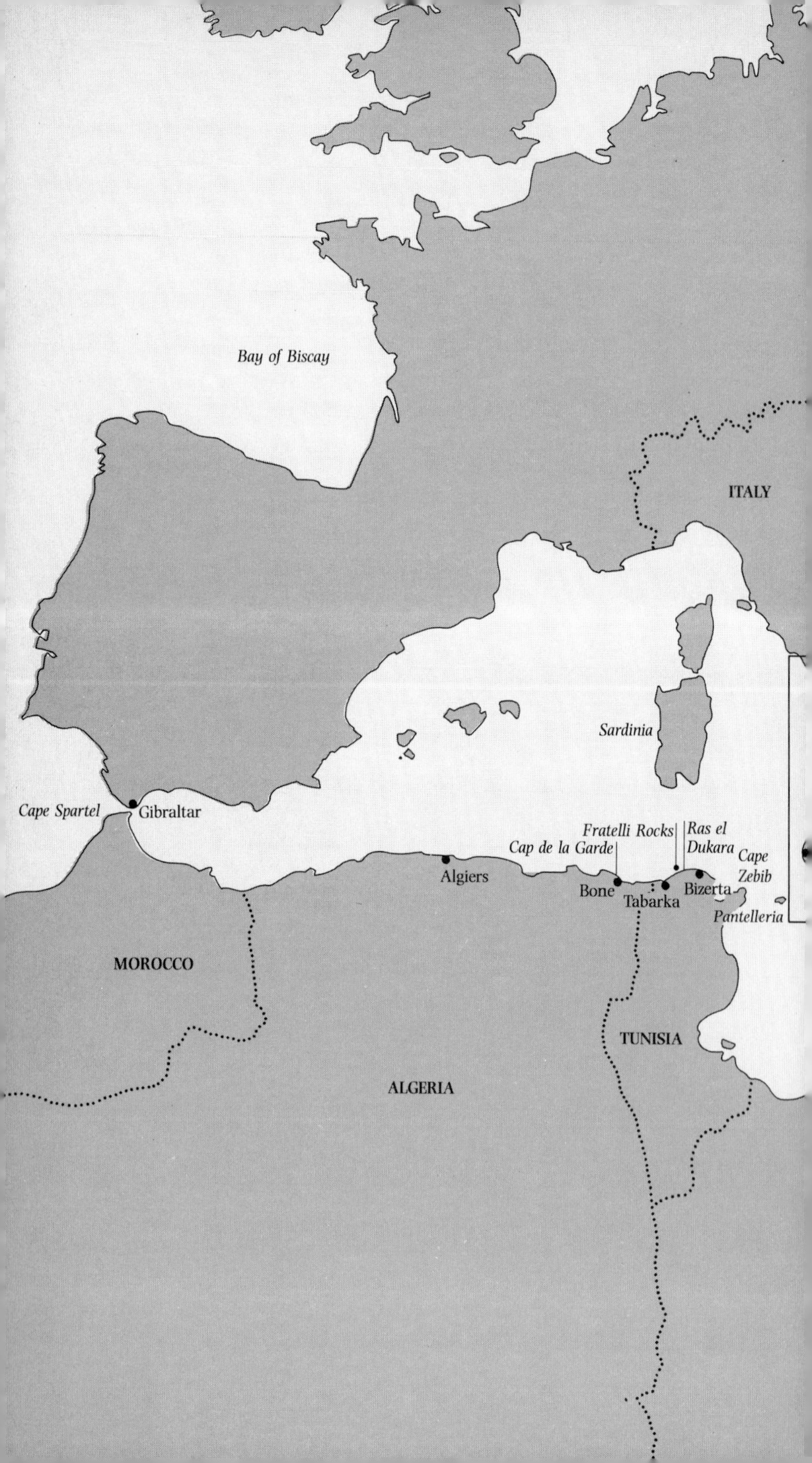

Bay of Biscay
ITALY
Sardinia
Cape Spartel
Gibraltar
Fratelli Rocks
Ras el
Dukara
Cap de la Garde
Cape
Zebib
Algiers
Bone
Tabarka
Bizerta
Pantelleria
MOROCCO
TUNISIA
ALGERIA

YUGOSLAVIA
Rogoznica
Korcula
Vis
See map
next page
Aegean Sea
TURKEY
Vathi
Samos
Kalimnos
Gulf of Mandalya
Leros
Kos
Simi
Dodecanese Islands
Rhodes
Kastellorizo
CYPRUS
Paphos
Beirut
Mediterranean Sea
Benghazi
Tobruk
Alexandria
Cairo
EGYPT

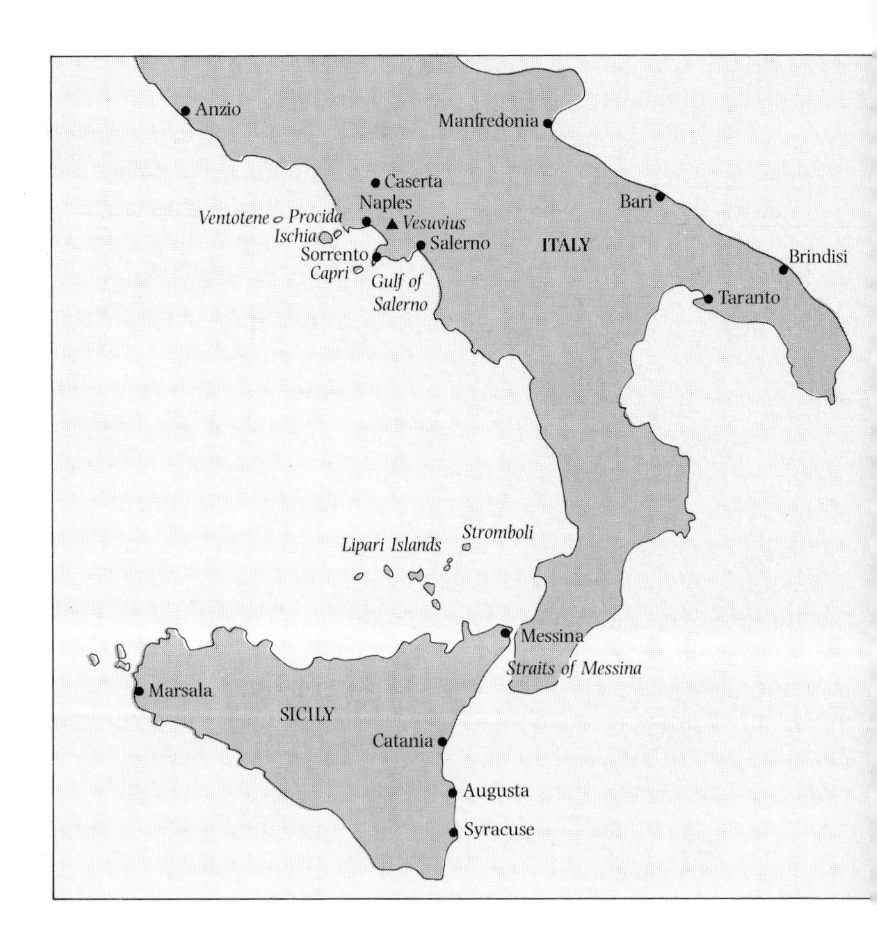

Southern Italy.

Algiers, where we thankfully got rid of the wretched tanks and checked our engines and equipment, aided this time by efficient base staff. As soon as this work was completed we made a high-speed hop of two hundred and eighty miles to Bone. We found that the harbour was littered with sunken ships and the port installations were a shambles of bomb-blasted quays and buildings.

Bone, which had been occupied by the Allies five months before, was still the most advanced port we held and supplies for our armies and air forces kept pouring through it. It was being bombed nearly every night. It was here that

Previous page: The Mediterranean

we encountered that very unpleasant Italian invention, the circling torpedo. These were dropped into the harbour by aircraft and, as in the old limerick, whizzed round and round in ever-increasing circles until they hit something.

We presented ourselves at the Coastal Force base, which seemed to have reasonable facilities and an adequate staff for servicing our ships. But the most important facility we required was non-existent: there were no fuelling pumps or fuel storage tanks. We soon found out what a strenuous chore it was refuelling from four-gallon petrol cans in the hot African sun after a night on operations in enemy waters. It was also terribly time-consuming, as our usual requirement was about four thousand gallons, which meant that a thousand cans, each weighing about forty pounds, had to be handled. This wasn't funny, especially as a very tired ship's company also had to clean and reammunition all the guns and carry out maintenance routines on engines and other sophisticated equipment.

What dismayed us most was that there was no trace of any of the multitudinous range of spares for our engines, guns and electrics. Mickey and Bobby Craven had been most painstaking in procuring and arranging shipment of a gamut of spares needed to keep our complicated ships in operation. They had even acquired a number of complete spare engines.

These shortages were to prove disastrous for our proud 19th MGB Flotilla, and in the ensuing months most of our ships were unserviceable for much of the time. Mickey as our Senior Officer and Bobby Craven as our Engineer Officer were unjustly blamed for inefficiency, which had a detrimental effect on their careers. Somehow Dicky Bird and I managed to keep MGB 643 consistently in service, with the result that we become the mainstay of the flotilla and went out on operations nearly every night. Some months later, when preparations were being made to move the base to Sicily, the missing cases containing our spares were discovered, hidden under piles of junk. They had been there all the time.

On 8 April, the day after our arrival in Bone, three of

our MGBS were pronounced operationally ready for action. They were Mickey Thorpe's 644, Tufty Forbes's 646 and my own 643. We set out that afternoon on the 130-mile run to Bizerta, which was the enemy's main stronghold in North Africa and also his principal supply port and air base. Although we spent many tense hours cruising round the approaches to Bizerta on that first night of our operations in the Mediterranean, we were disappointed to make no sightings of enemy vessels. We cleared Bizerta Bay and headed for Bone just before first light. When we tried to call up the base we found that we were in one of those peculiar 'dead' areas for wireless transmission and reception.

On our return we discussed this with the base staff, who told us that they were already aware of this phenomenon. They had in fact obtained some carrier pigeons and asked us to experiment with them on our next Bizerta run, which took place that night. We entrusted the pigeons to Tufty. Not knowing anything about this sort of job, he made a hopeless mess of it when he released the unfortunate birds with various messages including our estimated time of arrival. Instead of writing on cigarette papers, he used whole sheets of a standard naval message pad. The poor pigeons were so loaded down with these heavy sheets that they had difficulty in getting airborne. They must have had to rest several times on the way home, because they arrived some hours after we did and perched, utterly exhausted, on the roof struts of the main shed in the base. Like many another good yarn, the story of Tufty's pigeons became wildly exaggerated. The dramatic but untrue ending was that the base staff had to shoot the pigeons down from the rafters to retrieve Tufty's messages. The taunts he received for weeks afterwards used to drive him into a fury, and nothing was more effective than miming the part of being an exhausted pigeon when passing his ship.

We continued to make the Bizerta run in MGB 643 every night, leaving in mid-afternoon to enable us to spend the maximum number of hours in enemy waters. It was important to conserve fuel and it was our practice not to

Hanging out the washing on board an MGB.

drive our ships at or near top speed except during an action or in an emergency. At maximum permissible revolutions our petrol consumption was quite frightening – two gallons per minute per engine, making a total of nearly five hundred gallons per hour.

When we returned to Bone on 14 April after the sixth consecutive fruitless trip we were all very tired and fed up with our lack of success. The only mild excitement we had experienced had been a couple of nights before, when we had sighted two cargo ships in the distance; we had tried to close them but they had mysteriously disappeared near Cape Zebib. It was small comfort to be told that even if our nightly excursions were only deterring ships from moving in or out of Bizerta our efforts were well worthwhile. A message awaited me at the base from Mickey Thorpe, and I was soon in conference with him and the redoubtable Lieutenant P F S Gould DSC and Bar. Gould had a brilliant record of successful engagements fought in the North Sea and English Channel when in command of the small Vosper type MTBs. This man seemed to attract violent action as readily as a magnet attracts iron filings. I was told that within the past year he had lost four or five first lieutenants who had been killed or seriously wounded while standing alongside him on the tiny bridge of his Vosper. He was now in command of the 32nd MTB Flotilla, the vanguard of which had sailed to the Mediterranean shortly before we did.

Mickey did not beat about the bush. 'I'm sorry that I have to send you out again tonight, Geoff, but there's an important job on and 643 is the only MGB available.' He and Gould then proceeded to brief me on the proposed operations, which I found unusually interesting. A report had come in through intelligence, stating that there were two enemy cargo ships stranded under the cliffs of Cape Zebib. It had been decided to send in a force consisting of MTBs 634 and 638 and my own MGB 643 under Gould's command to destroy them. On the way back this unit was to move close inshore near Ras-el-Dukara, as the army had requested a diversionary bombardment from seaward on enemy coastal positions there. As I walked back to 643 I couldn't help thinking that these two ships could well be the ones we had sighted. In their panic they could have run aground but they would feel quite safe under the protection of the Zebib guns, which included the largest in North Africa with a calibre of 9.2 inches. No one was likely to pick an argument with those.

By midnight we were in position off Cape Zebib. It was a lovely calm night, but it didn't stay that way for long. Within minutes we went into the attack with engines roaring and guns blazing. We got so close under the cliffs that the enemy couldn't depress their big guns sufficiently to bear on us, and we became engaged in a series of actions which were the first ever fought by 'D' boats in the Mediterranean. The velvety darkness was abruptly torn apart and lit up by gunflashes, star-shell and searchlights from Cape Zebib and by the luminous hosepiping streams of tracer shell and bullets from our three ships. It was as spectacular as a large-scale public fireworks display. The noise was shattering. The staccato barking of our rapid-firing guns contrasted sharply with the deep slow booming of the Zebib battery; and all the time in the background came the heartening roar of our twelve supercharged Packard-Merlins pouring out of open exhausts as they ran at maximum revolutions.

The first torpedo was fired by Gould. It was defective and didn't run true, exploding on the beach with a blinding flash. The bang must have been heard for miles. Gould's second torpedo, fired almost immediately afterwards, hit the target amidships with another tremendous explosion. That cargo ship must have been a complete write-off and the other one must have been severely damaged with all the gunfire we poured into her.

As we disengaged we saw flares and turned to investigate. Searchlights groped for us and revealed two enemy submarines a couple of miles to seaward of us. Once again we raced in to attack – Gould and myself with gunfire and depth charges, and 'Shortie' Eason in MTB 634 with torpedoes. These both missed, passing close ahead of the leading sub. We lost both submarines when they apparently crash-dived. It would have been folly to hang around in that hot spot to find out if we had done them any lasting harm.

The searchlights and battery were still active when we made another sighting; this time it was two enemy destroyers. They wanted nothing to do with us and made off at top speed, not knowing that we had no torpedoes left. However, they let us have several parting shots in the form of a salvo of 4-inch shells, none of which found its target. We chased them for a while, but it soon became obvious that we had no hope of catching them. This was probably just as well for their armament was vastly superior to ours.

In any case it was time to keep our date with the army. Once clear of Bizerta Bay, we closed each other for a quick conference, which established that no one had been hurt and that none of our ships had suffered any serious damage. We put on another splendid pyrotechnic display when we carried out quite a lengthy bombardment near Ras-el-Dukara.

We were safely back in Bone by 8.45am. We were in high spirits and it was evident that this baptism of fire had given us great confidence in each other and had further strengthened our team spirit. I was proud of the coolness my crew had displayed. I was particularly impressed by the behaviour of two of our original hard cases – gunnery ratings manning the two-pounder pom-pom on the fore-deck. This gun had jammed during our bombardment of the cargo ships. As we disengaged and came under fire from the shore I shouted to them. 'Lie flat on the deck!' They ignored my order, climbed on top of the gun and shouted back in broad cockney, 'We'll get the bleedin' breech block fixed while we got light from their fuckin' star-shells.' And so they did, actually removing the breech block to clear the jam and replacing it within a couple of minutes.

After the Cape Zebib episode, Mickey decided that, come what may, MGB 643 must be taken out of service for a few days to rest the crew and for much needed maintenance to be carried out before we went phut like the rest of the flotilla. While the general shortage of 'D' boat spares was serious enough, we had an additional problem in that we were the only Coastal Force flotilla in the Mediterranean at that time equipped for electric cooking. The power for this was provided by two separate generators, driven by standard Ford small-car engines which had never been intended for the continuous heavy duty service to which we had to subject them; and they were simply not up to the job. In Britain the circumstances were quite different, because the Coastal Force bases there were equipped with quayside plug-in points and the generators only had to be run when we were at sea. Not surprisingly, no such facilities existed at Bone, nor at most of the other forward bases we came to use in the Mediterranean. Faulty generators were the Achilles heel of our flotilla and accounted for much of the unserviceability that occurred.

Another major headache was the lack of special hydraulic fluid for our power-mounted guns: the substitutes we tried could not be relied upon. Nothing was more disconcerting and discouraging than to find, while trying to fight an action, that a particular gun couldn't be trained or elevated properly.

During the last two weeks of April and the first week of May 1943 numerous actions were fought by the ships of our Coastal Force base at Bone. The resulting casualties included a number of friends. Stewart Gould was in the thick of it again. On the night of 25 April he sank two Siebel ferries as they entered Bizerta Bay, crowded with German troop reinforcements transported from Sicily. The Siebels were known to be well armed, and Gould was puzzled that they hadn't put up any fight – until he picked up some survivors who told him that his MTBS were thought to be German E-boats sent out to escort them into Bizerta.

Two days later Gould lost another of his officers in an incredibly courageous daylight attack on a big German supply ship escorted by two destroyers and an umbrella of thirty aircraft. He also lost about half of his ship's company, his ship and his own life. It was rumoured later that although he had been recommended for a posthumous VC the award had been prejudiced by the premature publication of the story of this magnificent exploit in *The Reader's Digest* under the title 'These, too, Were Expendable'. Whatever the truth of the matter was, we were disgusted when we heard that Gould had only received a Mention in Dispatches. In those days the extraordinary British system allowed for only one or other of these awards for gallantry, if made posthumously – the highest or the lowest. All the decorations for bravery or distinguished service between the two were strictly for the living. For those of us on active service in Coastal Forces in the Mediterranean in 1943, there could never be any doubt that the memory of this outstanding officer, who had served with such devotion and with complete disregard for his own life, ought to have been kept evergreen by the award of the highest honour his country could bestow.

 Ships of the 19th MGB Flotilla leaving Algiers, 7 April 1943.

MGB 643 continued to operate in enemy waters nearly every night, but we engaged in nothing spectacular. Then, early in May, we ran into trouble. There were just the two of us, with Tufty Forbes leading in his MGB 646 and my MGB 643 tucked close under his stern as we headed along the coast for Bizerta on a lovely sunny afternoon, when suddenly four Messerschmitt 109s pounced on us. Those boys were good. They made several low-level runs, attacking simultaneously from different directions and raking us with cannon shell. We scored some hits on them, but nothing good enough to knock any of them down. The contest finished up with honours about even. Little damage was caused to our ships, although we had quite a number of wounded who needed prompt medical attention. We decided to make for the nearest harbour, which was Tabarka – once a lair of the notorious Moorish pirates. We radioed for a medical team and ambulances to meet us there.

Most of the wounded were in 646, but we had a few. I was particularly concerned for Basil Bourne, who had been hit while standing alongside me on the bridge of 643. His own ship was unserviceable again and he had kindly volunteered to sail with us to help out. I was relieved to hear that his wounds, though painful, were not thought to be too serious. He was taken below to be cared for in my cabin, and I went to see him as soon as we were berthed in Tabarka. He had fairly large chunks of cannon shell in both his buttocks and was lying belly downwards on my bunk. I poured out two stiff whiskies. We laughed as we realised that there was no way Basil could swallow his drink in a prone position. I had to roll him gently on his side and hold his head up.

It seemed a long time before the ambulances and medical team came. Meanwhile Bone base radioed instructions for us to return there to reammunition and to take aboard some new crew members to replace our wounded. MGB 643 was on active duty again by the following afternoon. By 7 May she had set up a record by carrying out ten consecutive operations and we were greatly relieved when we re-entered Bone that morning to be told that we were to be withdrawn from service for a couple of days and 'rested' while we carried out maintenance routines with the help of the base staff.

I realised that the only way I would be able to get a proper rest would be to escape from my own ship. As soon as I had put everything in hand I paid a visit to a destroyer, which was berthed on the other side of the harbour for repairs. I knew her captain slightly and when I explained my situation to him he was very sympathetic. He placed an officer's cabin at my disposal, where I was able to enjoy a marvellous uninterrupted sleep for seven or eight hours. I dined in the wardroom and spent a most relaxed and enjoyable evening with my new-found friends. Although I was invited to stay on board overnight I thought it would be wise to return to my own ship. I had a two-mile walk round the harbour to reach the Coastal Force base. When I approached 643 about midnight, I was startled to hear her engines running and the order, 'Let go for'ard. Let go aft.' I broke into a run and leapt on board as she moved away from the quay. I was furious and raced along the deck to the bridge shouting, 'What in hell is going on here?'

It was unbelievable to me that my ship was putting to sea under the command of my first lieutenant. Beside him stood a brass-hatted RN commander, who said, 'You must be Lieutenant Hobday. I'm Kimmins,[1] Naval Intelligence. Sorry about this, but we've just had news that Bizerta has fallen and Commodore Oliver[2] has already sailed the supply convoy. He's in MTB 637 ahead there. He's commandeered these two ships to take us to Bizerta in time to get things ready for the arrival of the convoy.'

The fall of Bizerta was wonderful news – the best we had had since I had been in North Africa. My anger changed to a feeling of positive exuberance; I knew how vital those supply convoys were to the army and the Royal

[1] Later Captain Anthony Kimmins (1901–1964). Chief of Naval Information, British Pacific Fleet, 1945–1946. Film producer and director, author and playwright.

[2] Later Admiral Sir Geoffrey Oliver (1898–1980). Commander-in-Chief, the Nore, 1953–1955.

Sunken vessel at Bizerta, May 1943.

Air Force, who relied on them for most of their needs. I also knew that the fall of Bizerta signalled complete and final victory for the Allies in North Africa. It couldn't be very long now before we would be mounting an invasion of Sicily and Italy.

Soon after daybreak on 8 May we were threading our way through minefields off Bizerta's breakwater harbour. The western entrance was completely blocked and when we tried the eastern channel we found that sunken ships barred all further progress. There was no sign of life anywhere and I had the eerie feeling of being in a city of the dead. We lay stopped for a couple of minutes with our engines running while we tried to size up the situation, which was not at all what we had expected. Then, suddenly and shockingly, all hell was let loose as enemy guns, great and small, concentrated their fire on us from almost point-blank range. They had lain doggo until we were held

An MGB making smoke.

securely in this trap, from which there seemed no escape and little chance of survival. We were sitting ducks, with no room for manoeuvre, hemmed in as we were by the block-ships and the harbour walls. The only escape route was commanded by machine gun nests on either side and by a battery of the deadly high-velocity 88mm guns. I thanked God that our engines were still running and quickly put 643 hard up against the harbour wall, to give room for Commodore Oliver's MTB to squeeze past as it turned and surged for the exit at full power. I got 643 turned short round with surprising speed, and headed after the MTB with throttles wide open to unleash maximum power; not one of our magnificent engines faltered until they were hit.

It has always amazed me how swift and instinctive human reactions to extreme emergency can be. They are not slowed down by the normal decision-making processes of the brain. As we shot through the harbour exit under a hail of shells and bullets I found myself giving orders to turn the ship to port and not to follow astern of the MTB to seaward. For a short time we were partly shielded from the incessant gunfire by the outer wall of the harbour, and, what was equally important, the turn took us up to windward of the MTB. We soon changed direction again, this time to starboard to take us out to sea on a parallel course with our consort, making smoke which drifted down astern of her and afforded her some cover.

Mercifully we were soon out of range of those death-spurting machine guns but the 88mm guns continued to harass us with their accurate fire as we zig-zagged madly to escape destruction. The muzzle velocity of this remarkable weapon was much greater than the speed of sound, so that the boom of the gun was not heard in the target area until after the shell had arrived. As soon as we were out of effective range I asked for casualty and damage reports. It was a miracle that no one had been killed and we had only two wounded; but three of our engines had been knocked out and only one remained functional, giving us a maximum speed of no more than 16 knots. MTB 637's speed had been reduced too, and she seemed to be in the same crippled state as 643. We had dozens of shell and bullet holes. Most of the shells had punched through our thin wooden hull without exploding, so our ship was still quite seaworthy.

As we neared the open sea and were congratulating ourselves on our deliverance, the Cape Zebib battery opened fire on us at long range with frightening accuracy. The first salvo was perfect for line but fell a little short, sending up great spouts of water about a hundred and fifty yards away to starboard. I knew the gunners would immediately elevate to correct their range and I decided to follow the old naval maxim, 'Steer for the last fall of shot', by turning the ship sharply to starboard. Sure enough, the next salvo fell directly on the spot where we would have been if we hadn't altered course. We continued to zig-zag, but for some reason Zebib became silent thereafter. Shell fragments from the two near misses knocked a few further holes in us, with a most unfortunate result. The porcelain bowl of our one and only officers' head was completely shattered, causing us great inconvenience and embarrassment during the weeks we had to wait for a replacement.

When at long last we were clear of Bizerta Bay and had set a westerly course for Bone, we thought that we had left our troubles behind us. But our ordeal was not yet over. Several Messerschmitt 109s swooped down out of the sun to complete the job that the guns had failed to do. I can remember muttering, 'Oh God! Haven't we had enough for one morning?' Then, as suddenly as they had arrived, the aircraft departed and left us in peace. This second reprieve was as inexplicable and as unexpected as the first, when the guns of Zebib went silent. It seemed as though we were destined to survive this day after all, and with that conviction in my heart I felt greatly cheered. Only one enemy remained – the sea itself. The wind had risen and it was blowing a moderate gale from the west, smack on our nose. The great foam-swirled waves slowed us down considerably and made conditions wretchedly uncomfortable for everyone, especially for the wounded. I began to think that the day would never end. By the time we entered the harbour at Bone at 5.30 that afternoon, I had been standing on the bridge for seventeen hours without a break.

On the quayside at Bone quite a crowd had gathered to welcome us back. Most of them were from the base staff and other Coastal Force craft; they had wondered if they would ever see us again. We assured them that at times we had had the same doubts ourselves. As well as the happy sight of our friends there was the sad one of ambulances waiting for our wounded. Our consort, MTB 637, had suffered much more seriously than we had. Her commanding officer, Lieutenant Smyth, and his first lieutenant, Sub-Lieutenant Ridler, had both received severe wounds and the second motor mechanic mortal ones while they ran the gauntlet of fire during their escape from Bizerta harbour. Both Smyth and his Number One were mown down on the bridge and the only remaining officer, young

Sub-Lieutenant Arundale, who had been Basil Bourne's navigating officer in MGB 645, had to take over command in the most difficult circumstances. He was inexperienced and only temporarily on loan to 637, but he coped so well that he was subsequently awarded a Mention in Dispatches.

Smyth, an enormous soft-spoken dark-haired Irishman, was indeed having a rough time. Only ten days before this affair he had commanded MTB 637 in that magnificent last action of Gould's and had braved a holocaust when he had taken her alongside Gould's stricken ship to rescue survivors. Among them was Stewart Gould himself, who died soon afterwards from his terrible wounds, his body riddled with machine gun bullets.

We were appalled when we discovered the origin of our tribulations on 8 May. A smart alec journalist, in the hope of making a scoop, had sent a message through to London claiming that Bizerta had fallen, when at the time the First Army had only penetrated to the outskirts. This false information was broadcast on the official BBC news at 9pm on 7 May. It was this story that had triggered Commodore Oliver's dash for Bizerta and caused our lives and our ships to be so needlessly put at risk.

Next morning Mickey called a meeting of 19th MGB Flotilla COs and suggested that as 8 May was a lucky day for us we fix that as the date for a flotilla reunion dinner in London after the war. Everyone agreed. The first reunion in 1946 was such a success that it became an annual event. Some years ago it was expanded to take in other 'D' boat flotillas and it now covers all MGB officers who served in the Mediterranean.

Later on I saw Commander Kimmins again. I had formed a high opinion of him the previous day when he had skilfully tended our wounded, and his sustained courage and cheerfulness had impressed us all. In my ignorance of the theatre, I didn't realise that he was a well-known playwright and author of that long-running West End success, *While Parents Sleep*. He had come down to the base to impart a number of interesting items of news. The First Army had forced the Germans to retreat from Bizerta only a few hours after our own hurried departure from there. The Eighth Army was on the move again too and was advancing rapidly through Tunisia after breaking the Mareth Line. Many prisoners had been taken and a large British naval force, which included several of our Coastal Force craft from Bone, was now preventing any possibility of a seaward escape by Rommel's Afrika Corps. Admiral Cunningham was determined that there would be no successful 'Dunkirk' for them, and had aptly designated this blockade Operation 'Retribution'. The final result was that over 250,000 German prisoners fell into our hands.

7

On 12 May 1943 the Germans laid down their arms and officially surrendered. The long-drawn-out Desert campaign had at last ended. Everywhere in North Africa, in all our fighting services, there were victory parades and great jubilation, and a directive was issued that all of us who had been on active service should be given local leave at the first convenient opportunity.

In 643 we needed a rest and a change from the cramped quarters in our ship. Bone was still the main Allied supply port and continued to be bombed nearly every night by aircraft operating from Sardinia and Sicily. It would be good to have a few nights away from the bombing and the terrible racket made by the tremendous concentration of anti-aircraft guns now in the town. We had become such a close-knit family that we unanimously decided to spend the first part of our leave together.

Our relations with the army were so good that we found no difficulty in borrowing all the camping gear we needed, as well as two trucks and a jeep. The entire ship's company set off in high spirits like a pack of kids going on a seaside holiday. It was perfect summer weather and the lovely coast to the west of Bone, studded with spectacular bays and beaches, was ideal for our purpose. About seven miles away, near Cap de Garde, we found just what we wanted – a delightful cove not too far from the road – and there we made camp. Inshore the water was crystal clear and further out it gradually became a deep azure blue.

We had a marvellous time for the first couple of days, but were then hit by a series of misfortunes. On the second night I was struck down by dysentery and on the third night we suffered damage to a few of our tents from bomb splinters. A group of heavy bombers coming in to attack Bone must have decided that discretion was the better part of valour when they sighted the great cone of fire from the anti-aircraft guns ahead, and unloaded their bombs near us as they turned for home. The following afternoon there was a shocking tragedy. One of the young seamen made an unauthorised attempt to drive the jeep, which went out of control and charged into a tent where another rating, Able Seaman Tansley, was enjoying a siesta. He was killed instantly in his sleep. That finished our holiday. We had to get the body to Bone without delay and report the details of the mishap to the authorities; and I needed medical attention for my dysentery. What a cruel turn of fate that this youngster, only eighteen or nineteen years old, should have survived so many hazards of war, only to be killed accidentally while on holiday by his own friend and shipmate.

It was a sad return to Bone for all of us. Because of the hot weather it was essential to have the funeral immediately. The shortage of timber in North Africa ruled out the provision of a coffin and so, under the direction of an army padré, our poor shipmate's body was sewn into a blanket. In spite of my illness, I attended the brief graveside service. After the committal, the padré took me to one side and said, 'In case you are not aware of it, I must tell you that your rating's next of kin will receive a bill of one pound from the army for the blanket he is buried in.' I was so horrified that I paid for it on the spot out of my own pocket.

Lieutenant AD McIlwraith.

May continued to be an eventful month for our flotilla. We didn't do much operationally, other than two investigatory patrols on the Sardinian coast. The emphasis was now on bringing our boats up to the highest state of efficiency possible under the circumstances. We lost Mickey Thorpe, who was appointed to command the *Bickerton*, a Captain class frigate completing at Boston Navy Yard, USA. We were sorry to see him depart. To our surprise, Lieutenant Alan McIlwraith, recently arrived from England, was promoted to lieutenant-commander and made Senior Officer, a job which we had anticipated would go to Tufty Forbes. 'Mac' was one of the quietest and calmest officers I ever encountered, and typically he smoked a pipe.

Dark-haired, with brown eyes and of average height, his leadership was derived from his capability and courage and our affection for him. Bobby Craven suffered from the stigma of the extremely bad serviceability record of our flotilla and was transferred to base engineering staff at Algiers. He was replaced by Lieutenant 'Pop' Perry, who was a good Engineer Officer but tended to be more conservative in his approach to problems than his predecessor. My excellent first lieutenant, Norman MacLeod, also moved on; but this was a self-inflicted wound as I had put in a strong recommendation that he should be appointed to command a 'D' boat.

A new flotilla of US Navy PT boats was attached to us for active service training, the first to operate on our side of the Atlantic. They were a fine bunch of officers and men and we got along very well together. Their Elco boats were similar in size, speed and armament to our Vosper MTBS. We were very grateful when, in the usual generous American way, they helped us as much as they could with spares and even gave us some hydraulic fluid for our guns.

With the advent of hot weather malaria had become rife and was causing more casualties among our services than enemy action. We were issued with mepacrine as a preventative and palliative. Thanks to German propaganda, it was widely believed that mepacrine would permanently damage a man's sexual powers. This was of course quite untrue but it made it almost impossible to ensure that everyone took his daily tablet – until we hit upon a simple solution to the problem. As each man came up to receive his rum issue under the supervision of an officer, he had first to place his mepacrine tablet in his mouth and then gulp down his tot of neat spirits. So although our complexions were turned sickly yellow by the drug, we in 643 stayed clear of malaria.

On 23 May we moved to Bizerta, which was to be the forward operational base for Coastal Forces. Army and navy port clearance parties faced a mammoth task there, having to deal with no fewer than twenty-nine sunken block-ships, as well as numerous mines and major repairs to sabotaged quays and cranes. HMS *Vienna* joined us to act as our mother ship, providing good cabin accommodation and such splendid facilities as bathrooms and a cinema. Before the war she had been one of the fast Harwich–Hook overnight ferries. None of us made any use of her cabins, but we did enjoy watching the occasional film show again. The *Vienna*, however, turned out to be a white elephant and became such a burden to us that we wondered whether in fact her intended role had become reversed and that we were acting as *her* mother ship. We were continually being asked to supply working parties for cleaning, painting and general maintenance. Worst of all, she was coal-fired. I'll never forget the day when we were asked to muck in with the *Vienna*'s crew to coal ship. When coaling in an emergency or under tropical conditions, it was a naval tradition that all officers and men toiled together – even the padré – on this filthy, back-breaking job. The temperature was just over 100 degrees in the shade and the work went on from dawn to dusk. We soon looked like an exceptionally shabby lot of nigger minstrels and it was a week or two before we became really clean again.

Almost immediately after our arrival in Bizerta we were once again under heavy operational pressure, with patrols being mounted most nights off the southern coast of Sicily and the island fortress of Pantelleria. Here for the first time we had the experience of being at the receiving end of radar-controlled gunfire. We used to say what a great navigational aid this was, because you could be sure that at precisely seven miles off Pantelleria the first salvo would arrive.

Our next really interesting assignment was the invasion of Pantelleria. This was to be essentially a navy job and we were briefed by Rear Admiral Rhoderick McGrigor,[1] a tiny quick-moving wisp of a man who was in charge of the operation, and by his chief of staff, Commander Geoffrey Robson.[2] In contrast to McGrigor, Robson was tall and

[1] Later Admiral of the Fleet Sir Rhoderick McGrigor (1893–1959). First Sea Lord and Chief of Naval Staff, 1951–1955.

[2] Later Vice-Admiral Sir Geoffrey Robson (b.1902). Lieutenant-Governor and Commander-in-Chief of Guernsey, 1958–1964.

HMS *Vienna* with MLs moored alongside.

powerfully built with a commanding personality. A large force of 'D' type MGBs was to be deployed in the Sicilian Channel north of Pantelleria to act as an outer screen in case of a sea attack by the enemy. We sailed on 10 June to arrive on station just before dusk. 643 was in the van, with McIlwraith, who was in command of the unit, on board. We were no sooner on our patrol line than we were attacked by six fighter-bombers. There was no mistaking what they were – twin-engined RAF Bostons. Although we repeatedly fired our Schermuly recognition signals, the RAF pilots were in no way deterred from making several low-level runs on us, letting us have all they'd got. Tempted though we were to protect ourselves, we didn't fire a shot but took violent avoiding action. None of the Bostons' bombs hit us and their cannon shell did us no serious harm. Our casualties were light – one killed and two wounded – but this was regrettable enough. I would have loved to listen in to the pilots' reports on this attack and hear what sinkings and damage they claimed.

Our orders were quite explicit. We were to maintain complete radio silence unless we sighted or engaged the enemy and we were to stay in our patrol area until we received a recall signal. We heard nothing all next day and saw nothing. It was nearing midnight when I said to McIlwraith, who was a much more patient man than I was, 'I'm getting fed up with this – you would think the silly so-and-sos would let us know what in hell's going on.' Mac just grunted, but my navigator had a bright thought, 'What say we tune in to the BBC, sir? They're bound to be telling the great British public all about it.' It was a damned good suggestion and when I acted on it we heard that the invasion had been a complete success and that Pantelleria had surrendered at 2pm – ten hours earlier.

Soon after dawn on our way back to Bizerta I sighted a large yacht under full sail. By this time, apart from the odd nap in the chart room, I had been on watch for forty hours and my strained eyes could no longer be trusted. I said quietly to the bridge lookout, 'Do you see anything bearing red 10?' He looked through his binoculars, lowered them, wiped them and looked again, and then in a voice of total disbelief replied, 'Looks like a big pleasure yacht, sir.' We closed it rapidly with all our guns trained on it. There were eight men in the cockpit, two wearing brass hats. 'Good morning, gentlemen,' I said. 'I think you'd better come on board – there's bad weather building up.' One of them thanked me, said they would be all right and were enjoying their cruise. His English was perfect, although he was obviously an Italian. 'I'm about to sink your ship,' I went on, 'so you'd better get aboard at once unless you want to go down with her.' They clambered on to 643 pronto.

Being a keen yachtsman myself, I found it a most distasteful task to have to sink this lovely thoroughbred by gunfire, but I had no option. She couldn't be towed through a war zone, nor left adrift to become a menace to navigation. As soon as we were under way again, Mac and I started to question our prisoners. The senior officer was

Captain Renato Pennetti, naval commandant of Pantelleria, and the others were members of his staff, so we thought we had made quite a valuable haul. Pennetti asked me where we were heading. I replied, 'It's none of your business,' and then thought for a moment and added, 'but I'd be interested to know why you have asked.' He smiled. 'It's just that I'd prefer that you don't run into any of our minefields, and I'm sure you must feel the same.' I took him down to the chart room and got him to mark the Italian minefields on the Tunisian coast. They were extensive and many of our ships were engaged in sweeping them. I made a signal to Admiral Dickens, the naval officer in charge of Bizerta, that we were bringing in an important prisoner who could give full information on these minefields.

Pennetti told me about how he and his companions had slipped away in the yacht during the night. He had prepared his escape when the prospect of an invasion seemed imminent. He was making for Sicily when we intercepted them. He also told me he would gladly assist our people in any way he could. He hated the Germans, who pulled out of Pantelleria soon after their surrender in North Africa, stripping the place of guns, ammunition and equipment and leaving their allies to carry the can. When we arrived at Bizerta there was quite a reception committee to interrogate the Italians, including naval intelligence and minesweeping officers. We were told later that by bringing Pennetti and his staff in, we had made a major contribution to speeding up the minesweeping effort and to reducing its hazards.

Almost immediately after the Pantelleria episode, we became involved in a very peculiar affair – so hush-hush that we never found out what it was really all about. The only one who knew was McIlwraith, who led us on these clandestine expeditions but disclosed nothing more than he had to, impressing on us the necessity to maintain complete secrecy. What he did tell us was that we would be making a number of trips to Sicily to land agents and glean information for our intelligence services.

On our first trip we towed a number of rather decrepit fishing boats and hove to about three miles off the Sicilian coast. One of them went in, to return shortly with the good news that the coast was clear. The rest of our passengers then disembarked and we parted company. They were a villainous-looking lot and, although they were supposed to be fishermen, we were convinced that they were a smuggling gang who habitually operated between Tunisia and Sicily.

On our subsequent trips we were always met by one of the fishing boats, with whose occupants Mac would hold secret palaver; sometimes he would go ashore with them. We were beginning to think that our agents had connections with the infamous and powerful Mafia. However that may be, there was far less resistance when the Allied landings took place than had been generally anticipated and a number of the shore batteries did not open fire. We suspected that it was not without significance when McIlwraith was awarded the French Croix de Guerre avec Palmes soon afterwards.

Meanwhile there was tremendous activity ashore and afloat. Large numbers of vessels of all types kept arriving at Bizerta, mainly 10,000–ton liberty supply ships and landing and assault craft. We were witnessing the build-up for the greatest amphibious operation up to that time – the invasion of Sicily.

On 25 June 1943 I received a briefing that filled me with such foreboding that I immediately wrote a long letter to Helen, covering everything I could think of that could be helpful to her as a widow. I left this with the captain of the *Vienna* to be posted if I did not return the next day. It was the only time I ever did such a thing. I had been given an apparently hopeless task. My instructions were to lead a unit of 'D' boats (MGBs 643 and 644, and MTB 651) through the fortified entrance of Marsala harbour on the west coast of Sicily to bombard the military aerodrome there. It seemed to me that even if we penetrated the entrance unscathed there could be no possibility of ever getting out again.

Not long before we sailed the chief motor mechanic of MGB 644 went down with dysentery and had to be put aboard the *Vienna*. Norman Macleod, who had taken over Mickey Thorpe's boat, asked me for help as his engine room staff was a pretty weak one and I impulsively lent him my incomparable Dicky Bird – a decision that I still regret to this day.

By 1am we were only a mile and a half from Marsala, sneaking slowly along in line ahead with silenced engines trying to make the entrance unobserved. Then, with jarring suddenness, there was a series of heavy explosions, which I thought at first was gunfire. To our horror, we discovered that we were surrounded by a large field of mines and all three ships had set some off. Instead of the horns of the usual contact mine, these loathsome things had slender antennae, stretching about 20 feet in every direction like gigantic and deadly spiders.

Our three ships lay stopped within easy hailing distance of each other while we discussed the situation and sorted ourselves out. Luckily, 643 and MTB 651 had no casualties and had suffered little damage, but 644 was in a very bad way – in a sinking condition with a fire in the engine room and a number of dead, wounded and missing. There was obviously no future in proceeding further into that impenetrable minefield. Our most urgent job was to help the crew of our stricken consort. We lowered our scramble nets and launched our dinghies and Carley rafts. We took off all the living, including the wounded, and searched for the missing in the water. It was thought that some of them had been blown off the deck at the time of the main explosion, and we did find two or three men, among them the coxswain, who had dreadful injuries. The engine room had become an inferno – and Dicky Bird was down there. I prayed that he had been killed outright at the moment the mine exploded. I helped to carry the coxswain on board 643 and made him as comfortable as I could on the bridge: he couldn't be moved further. His back seemed to have been broken and he had two fractured legs, which we set and splinted. He was delirious with pain, so I gave him an injection from an ampoule of morphine.[1] We continued our grisly search until just before dawn, in the uncomfortable knowledge that we might set off another mine at any time; but we found no more survivors. I was broken-hearted over the loss of Dicky Bird, whom I still remember with respect and affection.

Before heading back for Bizerta we had the unpleasant task of ensuring that nothing remained of our unfortunate sister ship to fall into enemy hands, and we poured shells into her charred hull. As we anticipated, it wasn't long before enemy aircraft came searching for us – three Stuka dive-bombers. As they prepared to attack, I disregarded one of the basic rules of gunnery and opened fire before they were in effective range, in the rather forlorn hope that an aggressive display might deter them. I knew full well that, once committed, they wouldn't pull out of their dive until they had released their bombs. To my surprise and relief, this ruse worked and they hesitated. We had gained the initiative. After messing around uncertainly for a while they turned back to Sicily.

As we were leaving harbour on our next patrol there was an embarrassing incident. I was leading a mixed unit, consisting of three of our 'D' boats and two PT boats, which had been attached to us for 'action experience'. The Americans were making a propaganda film showing the close cooperation between United States Navy and Royal Navy ships in the Mediterranean war zone, so there was a camera crew on the stern of 643 taking shots as we sped in close formation in line ahead. At the request of the cameramen we were travelling faster than we usually did in harbour and must have made a brave sight, but not alas, for long. I handed over to Holway, while I went aft to talk to the film unit. A few minutes later we hit an underwater obstruction with such force that we hurdled over it like a Grand National steeplechaser. One of the PT boats had no chance of taking avoiding action and repeated our inglorious performance, but fortunately the other three ships

[1] Author's note. I am glad to say that the coxswain recovered – I bumped into him at Waterloo Station later in the war.

were able to sheer off. We had charged over a sunken breakwater that was quite clearly marked on the chart and furthermore it was buoyed at the seaward end. Young Holway must have been dreaming. He was quite familiar with this hazard but had taken the ship inside the buoy instead of to seaward of it. It was not long before the breakwater was being referred to as 'Hobday's Folly' and many of my colleagues marked their charts accordingly.

We had damaged all four of our propellers and bent two shafts. Instead of displaying our efficiency and local knowledge we had managed to do just the opposite. The whole thing was an absolute bloody débâcle, and as we limped slowly over to the nearest berth with much juddering and vibration I was smarting with chagrin and shame. It only needed the personal signal from the Commander-in-Chief, Mediterranean, Admiral Sir Andrew Cunningham, to complete my misery. It read: 'You have incurred my grave displeasure consequent upon the grounding and damage to MGB 643 and the American PT boat.'

I was in a really parlous situation. We had no replacement propellers or shafts and there would be no slipping or docking facilities available to us for some time. They were all booked for ships being made ready for the invasion of Sicily. Even when it came to filling in the official 'Collision or Grounding Report' form, I had no room for manoeuvre or soft pedalling – everything that had happened was recorded on film in glorious Technicolor.

However, our luck soon started to turn. It was pure chance that had led us to berth where we were – alongside a revetment in about eight feet of perfectly clear water, with a hard clean sandy bottom. We were drawing about six feet, and couldn't have been better placed for underwater work. There was a big ex-French submarine base nearby which had been taken over by the Americans, mainly because of its extensive, well-equipped engineering workshops. There was nothing we could hope to do through official channels but, having always believed that 'God helps those who help themselves', I made a personal approach to the American CO. As always the Americans were friendly and helpful. My object was to borrow a diver and a mobile crane, but they could spare neither at the time. When I explained what I wanted them for they offered to lend me three shallow-water diving kits and some first-class chain blocks and tackles. 'I don't think you've got a cat's chance in hell of drawing your Gawd-damned shafts with that toy gear,' said the CO, 'but if you guys get them out we'll fix 'em real good.'

He sent a chief petty officer artificer back with me in a truck, together with all our scrounged equipment. The artificer had a good look at the job and demonstrated the use of the diving kits by going down and inspecting the propellers, which he assured us could be straightened out. He agreed to make some tapered wooden plugs to seal the stern glands as we drew the shafts. 'I'd better send you down a mobile fire pump to keep your bilges dry,' he suggested. 'Otherwise if anything goes wrong when you are drawing the shafts you boys will be in big trouble.'

We were now so fired with enthusiasm that we were determined to carry out this tricky work with the utmost speed. The worst part was the underwater operation, in which we nearly all participated and which we found most exhausting. In fact ten-minute spells were long enough. The diving kits, which were of a new pattern, comprised only a helmet, air hose and signal cord, and the air supply was provided by a large automobile hand-pump. We always felt insecure down below, as the kits were only designed for brief inspection dives, and we experienced a considerable sense of relief each time we came to the surface.

We rigged the blocks and tackles over the transom, with loose wire strops round each shaft, and the engine room staff followed through each shaft from inside with the tapered sealing plugs as they were drawn. Each shaft with propeller was almost a ton in weight but quite a bit less under water. Everything went smoothly and we had no accidents. The Americans were greatly impressed by our efforts and managed to find a mobile crane to lift the shafts off the sea bed on to a waiting truck. They sent the crane back again with the repaired shafts and props.

Much to the surprise of everyone concerned, we were ready for operational service again within forty-eight hours. What was important to us was that we had not only carried out extraordinary repairs to our ship but we had also retrieved, and indeed enhanced, our good name, and had fully restored our morale.

8

It was blowing hard and there was a nasty sea as we plunged northwards towards the Straits of Messina on the afternoon of 9 July 1943. Our three little ships, all 'D' type motor gunboats, must have made a stirring sight, running in line ahead with battle ensigns flying and the steel-helmeted gun crews closed up at full action stations. Our Senior Officer, Alan McIlwraith, had decided to lead the unit in MGB 643. Tufty Forbes commanded MGB 646 and MGB 641 was under the command of the imperturbable and good-natured Peter Hughes, who had been a district officer in Basutoland before the war. Our orders were most unusual. We were to enter the straits, one of the most heavily defended areas in Europe, before dusk 'so that the enemy can clearly observe and identify you'. Furthermore, if any of us was hit by the shore batteries that abounded on both sides of the straits, we were instructed that the surviving ships were to sheer off behind a smoke screen. 'Under no circumstances are you to attempt to aid a stricken ship or to rescue her crew.'

We had been carefully briefed two days before, when we were told that Operation 'Husky', the Allied invasion of Sicily, was timed for first light on 10 July; over three thousand ships and landing craft would descend on the southern corner of the island. One of the Allied fears was that the two halves of the large modern Italian fleet based at Naples and Taranto might converge on the straits during the night of 9 July and create havoc among the armada of landing craft. During the planning sessions the Allied Air Forces stated that while they could prevent such a conjunction during daylight hours, only the navy could hold the straits during the night, and would have to continue to hold them until the conquest of Sicily had been completed. It was thought that if the enemy knew that the Royal Navy was actually operating in the straits they might be deterred from attempting to interfere with the landings. This made sense to us, and so did the 'no rescue' order, much as this went against the grain. It was quite clear that if one of our ships was stopped by a direct hit from a shore battery, the enemy would have the exact range and bearing to destroy any other ship coming alongside to give aid.

Because of the unfavourable weather many thought that the operation would have to be postponed. They included the Germans and Italians in Sicily who had been standing to in anticipation of an invasion, but on the night of 9 July must have heaved a sigh of relief and taken the opportunity to catch up on some sleep. The Axis navies ceased shore patrols and lay in harbour.

We were about two miles off the Calabrian shore when we sighted a big Italian submarine on the surface, with many of its crew on deck taking the air. We closed at maximum speed and managed to rake her with gunfire before she crash-dived and disappeared. We let go depth charges, but a nearby shore battery opened fire and gave us no chance to search for evidence of the success of our attack such as an oil slick. The shore battery's reaction was an undoubted sign of our success in getting ourselves identified as British. The shooting of the battery was so good that a shell from their third salvo scored a direct hit on MGB 641 amidships, and it was obvious that there was no hope for her. Tufty and I complied with our hateful orders and abandoned our friends, suffering a terrible feeling of shame and anger. Not long afterwards, when darkness had started to enshroud us, I said, 'I don't like this, Mac.' 'Nor do I,' he replied, 'but what the hell can we do?' I signalled to Tufty to come alongside 643. I shouted, 'We just can't abandon Peter and his boys. What do you think?' Tufty was fairly bristling with rage. 'Of course we're bloody well going back. If we hadn't stopped now, I'd have disobeyed orders and gone back on my own.' Mac entirely agreed and we worked out a plan. As soon as we located Peter's boat, 643 would position herself between her and the battery, laying a smoke screen for exactly ten minutes while Tufty went alongside and, if she was still afloat, took all the living off the ship or hunted for survivors among the wreckage.

We went about our search very quietly, making as little noise and wash as possible, and were apparently unobserved by the battery. I don't suppose they thought we would be crazy enough to return. Nevertheless in 643 we were glad when our ten-minute deadline had expired. To our dismay we found Tufty still alongside 641, which remained at least half afloat. He said he needed another eight or ten minutes, so we had to go back and make smoke again. At last, to our relief, we were able to get away from the shore battery and resume our patrol up the straits. There were no further excitements that night, and we were thankful to find that Peter Hughes had miraculously suffered no casualties.[1]

As dawn approached we headed our ships for Malta, a hundred and fifty miles away. We shaped our course so as to keep well clear of the landing beaches. Soon after daybreak we witnessed an awe-inspiring and heart-lifting spectacle – the greatest concentration of ships that human eyes had ever beheld. Overhead the sky was heavy with the sight and sound of innumerable aircraft. The enemy had been taken completely by surprise when our landing craft swept on to their beaches and discharged their lethal cargoes. In those first vital minutes there was practically no opposition.

Once we had reported to our base in Malta, we held our own de-briefing session. We were curious to learn from Tufty what had kept him so long when he was making his

[1] According to Admiralty records, MGB 641 was sunk on the night of 14 July 1943.

rescue bid. He seemed very shifty and evasive about the whole episode and, although we often laughed about it afterwards, we were furious that day when we finally extracted the truth from him. We learned that once he had picked up all the survivors he had relieved the sinking ship of stores, equipment and provisions, and had even removed the barrels of the Oerlikon guns to replace his own badly-worn ones. While he was engaged in this pillaging he had undoubtedly endangered our ships and men. Peter Hughes was very grateful to us for what had been done and presented me with his service revolver, complete with webbing belt and holster, as a memento of the occasion. He would report his as lost with the ship and be issued with another one. I don't think Peter gave anything to Tufty, no doubt feeling that he was already overdrawn in Peter's bank of gratitude.

Next morning we departed from Malta to patrol the Straits of Messina again. Another 'D' MGB replaced Peter's ill-fated ship. Comparatively large numbers of Coastal Force craft were now involved in the Sicilian operations. MLS were carrying out escort and patrol duties, covering the continuous stream of landing craft bringing further troops and supplies; and 'D' type MGBS and MTBS, together with the high-speed 70-foot Vosper MTBS, guarded the Straits of Messina and their approaches every night. Coastal Forces were involved in several actions during the next few weeks in or near the straits, one of the most remarkable being that fought by Lieutenant Christopher Dreyer on the night of 13 July. MTBS 81, with Dreyer on board as Senior Officer, 77 and 84 were lying with engines cut in the narrows off the town of Messina when two submarines suddenly bore down on them. The range was too close to permit the MTBS' torpedoes to level off so they crash-started their engines and went full astern. MTB 81 was still going astern as one of the submarines crossed her bows. Dreyer fired his torpedoes and completely destroyed the submarine, which turned out to be U561.

Troop transports *en route* for Sicily, 10 July 1943.

Three nights later, on 16 July, Tufty and I were ordered to patrol the southern end of the Straits of Messina. We were told that Denis Jermain would be up there too in MTB 315 leading three other Vospers of his 10th MTB Flotilla. Well after midnight, to my astonishment and consternation I sighted a cruiser steaming towards us and felt the cold hand of fear grip my heart. The cruiser was heavily armed and it seemed suicidal to attack her. We would be destroyed long before our small guns could be in effective range and in any case the amount of damage we could inflict would only be minor as we didn't carry torpedoes. Even so, it would have been shameful to turn away. Our quandary was almost immediately resolved for us when Denis Jermain's four Vospers roared in to attack. We were stunned by the intensity of the gunfire which poured out of the cruiser and filled with admiration at the courage of Jermain and his companions. Three of the MTBs were hit; one of them became momentarily an incandescent ball of fire from which no one survived. The other two were not seriously damaged and all three remaining boats got their torpedoes away, but seemingly without effective result.[1] Aboard 643 we were greatly relieved when the cruiser altered course and disappeared into the night.

We learnt afterwards that she was the Italian light cruiser *Scipione Africano*, which had the fantastic speed of 41 knots. She was heavily armed, mounting eight 5.3-inch, five 40mm and eight 20mm guns.

Later that night we were in action against Italian MTBs and the German army. This was one of several brief skirmishes we had with the Italians, who invariably fought with great verve and gallantry. It was no discredit to them that, as on other occasions, they quickly disengaged when they discovered that they were up against a heavily armed 'D' MGB and used their superior speed to escape us. Any sustained engagement would have been fatal for them, and in any case their job was to fire their torpedoes at bigger and more valuable targets. They were commanded by men

[1] MTB 316 was sunk and MTBs 260 and 313 suffered superficial damage in this encounter.

of exceptional courage and skill, many of them from old Italian noble families.

After our clash with the Italians we stole into a bay where the main road from Messina to Catania bordered the coast, and spotted scores of laden supply lorries moving south. They were completely surprised when we opened fire and we were able to destroy several vehicles. We happened to have an American journalist on board, who had asked if he could accompany us on a mission in enemy waters. We thought that he would be overwhelmed by such an exciting night out. But he had been fed with a lot of highly coloured press reports and propaganda about MTBS and MGBS, and assumed that the exceptional patrol we had undertaken was just our normal routine.

Shortly after the initial landings in Sicily we had based ourselves first on Syracuse and then, after its capture on 13 July, the great Italian naval harbour of Augusta, where the enemy mounted extremely heavy air attacks every night. When we did have the odd night off from our patrols it was almost impossible to relax or get much sleep. When returning to Augusta after an operation we had to pick our way through all sorts of wreckage and debris produced by the night's bombing. Often there were grimmer relics – the floating bodies of seamen and airmen.

One morning in Augusta a colour sergeant of 41 Royal Marine Commando came aboard with a note from Lieutenant-Colonel Lumsden inviting me to lunch. This was the colonel I had become so friendly with in the Isle of Wight and to whom I had jokingly remarked, 'I'll see you in Sicily.' The colour sergeant had a jeep and we were soon driving northwards towards Catania, where the Germans had effectively blocked any further advance by Montgomery's Eighth Army. After a while the sergeant observed, 'There's only a few miles to go now, sir, and the colonel said to tell you that we are living a bit rough – he knew you wouldn't mind.' As we drew up, some German shells burst nearby. The sergeant said, 'We'd better run for it, sir. I'll lead the way,' and we dashed through an olive grove, arriving breathless in a slit trench where I was warmly greeted by Bertie Lumsden. Our alfresco meal was more informal than sumptuous, consisting of bully beef and baked beans. The shelling continued intermittently. 'Thought you'd be interested to see how the poor live,' said Bertie. I enjoyed meeting him again and we had many a laugh as we swapped news. He had his own particular brand of humour, of which this invitation to lunch in the front line was characteristic.

Towards the end of the Sicilian campaign we were ordered back to Bizerta to rest and refit in preparation for the forthcoming invasion of the Italian mainland. My crew had continued to acquit themselves well in Sicilian waters and 643 had an unbroken record of operational availability. This was largely due to Petty Officer Motor Mechanic Hunter, a likeable but somewhat taciturn man who had replaced our beloved Dicky Bird. Hunter's ability and devotion to his duties bid fair to rival the example set by the incomparable Dicky. By Coastal Force standards our new engine room chief was quite elderly, being about my own ripe old age of thirty-five. A head garage motor mechanic in peacetime, Hunter was one of those rare human beings who are almost completely unaffected by high voltage electric current and he could handle live wires that would have killed any ordinary mortal. He had a good steadying influence on the crew.

After a pleasant and peaceful passage from Augusta, morning found us about to enter Bizerta harbour once again. After our previous experiences there, I suppose it was not surprising that I felt distinctly uneasy as I wondered what fate held in store for us this time. However, all seemed well as we berthed in almost exactly the same spot we had used for our underwater propeller-changing feat. We had just finished squaring-off the upper deck and rigging a gangway when a naval dispatch rider appeared and handed me an impressive looking envelope. Inside was a formal invitation to join Vice-Admiral Sir Gerald Dickens for luncheon that very day. I barely had time to shower and change into clean white tropicals before a jeep arrived to take me to his villa. I was both flattered and puzzled by

The three oldest things aboard MGB 643.
The author, Chief Motor Mechanic Hunter, the after 6-pounder gun.

this invitation. I had never met the admiral, who was Flag Officer, Tunisia, but I knew something about him. He had retired shortly before the war, but had volunteered to serve his country again. I also knew he was popular and that he was the grandson of the novelist, Charles Dickens.[1]

Perhaps, I thought, the admiral is using this opportunity to thank me for sending in Captain Renato Pennetti and his colleagues as prisoners from Pantelleria. I turned to the driver, 'Is the admiral throwing a party then?' 'Yes, sir,' he replied, 'not a big one, but rather special. You'll see, sir,' he added as though to forestall further questions. When I was greeted by Sir Gerald at his villa he said, 'Ah, Hobday. So glad you could come. Rather a special little occasion today. We're entertaining some very charming young ladies,' and he took me through and introduced me to four

[1] Sir Gerald's son, Peter Dickens, had a distinguished record as Senior Officer of the 21st MTB Flotilla based in home waters.

Wren officers, the first to arrive in Bizerta. They looked fresh and lovely in their trim white uniforms, quite delicious in fact. Few of us had seen an English girl since leaving Milford Haven five months before, so this was a most delightful surprise. The other guests were a destroyer commander, a major of the Royal Engineers and the flag lieutenant. Conversation flowed freely over aperitifs.

Just before we sat down for lunch the admiral took me to one side and whispered, 'Have you got the crayfish?' I looked at him in astonishment. 'What crayfish, sir?' 'Haven't you just come in from Fratelli Rocks?' 'No, sir, I've just come down direct from Sicily.' His jaw dropped and he burst out with, 'My God! We've asked the wrong man to lunch!' Then, quickly recovering himself, he apologised for his inadvertent rudeness. He was nothing if not a gentleman. It was a splendid party and by far the most enjoyable experience I had had since my arrival in the Mediterranean. As we were leaving the admiral said, 'You must come and lunch with us again soon, Hobday,' which was no doubt his courteous way of making amends for his unfortunate remark. He was as good as his word, and it wasn't long before I was his guest for another agreeable luncheon party.

Bizerta was in some ways an ideal place for our rest and refit as there were no bombings and no diversions of any kind. The town was in ruins and most of the original inhabitants had disappeared. None of the normal pleasures of leave or leisure were available, not even the simple ones, such as shopping or dining out. On the other hand we had plenty of peace and quiet, which we badly needed, and for recreation we had unlimited swimming right by our ship. In the evenings we often exchanged visits with other ships, and this was the main antidote to an awful feeling of isolation and boredom.

Operation 'Avalanche', the amphibious invasion of the Italian mainland in the Gulf of Salerno, started on 9 September. By that time MGB 643 was ready for sea and was awaiting orders to join in the operation. The fact that we did not receive these orders until the following evening

Italian warships off the North African coast on their way to surrender, seen from the deck of HMS *Warspite*, 10 September 1943.

enabled us to witness one of the most historic events of the war – the surrender of the Italian fleet to the Royal Navy. Admiral Cunningham had given instructions that the Italian ships were to approach the North African coast at Bizerta and then proceed to Malta. We saw him leave Bizerta harbour, accompanied by General Eisenhower, in the destroyer HMS *Hambledon*. We were a little disappointed about this, because we had been alerted to be ready if required to take these two great men out to view the Italian fleet from 643. What an experience it was to see a column of some of the world's most modern warships, led by Cunningham's old flagship the 15-inch-gunned *Warspite*. It was a most moving spectacle – one I'll never forget – and it made me feel proud to be an officer in the Royal Navy.

9

I was still absorbed in the surrender of the Italian fleet when the arrival of our sailing orders jolted me back to concentration on our own immediate affairs. I was instructed to sail MGB 643 to the battle area at Salerno and report to Commodore Oliver aboard his headquarters ship HMS *Hilary* for further orders. We were to refuel and rest overnight at Messina. I had come to know Commodore Oliver when he was Senior Officer, Inshore Squadron, in North Africa and I respected him for his proven ability and valour. A stocky, thickset man, he was a seagoing bulldog if ever there was one.

Once again MGB 643 was the only serviceable ship in the 19th MGB Flotilla and I could not help feeling a bit lonely. I would have liked to be sailing in company with the sister ships of my flotilla in this latest incursion into enemy territory. It was some six hundred miles to Salerno and the first leg of the passage took over twenty hours at cruising speed. It was wonderful to be able to sail through the Straits of Messina without fear of attack. By the time we had laboriously refuelled from jerricans at Messina we were glad to find a berth and snatch a few hours' sleep before we turned to again at 4.30am. Just after dawn we were threading our way through the Lipari Islands, passing close to the shores of the great volcano of Stromboli, which although smoking was not in active eruption. It was now 12 September, the fourth day after the first landings at Salerno. The weather was perfect – there was no wind, not a cloud in the sky, and the deep blue sea was plate-glass smooth. The coastline was dramatic and with all this peace and beauty around us it was quite incongruous that we now had to be engaged in making our ship 'ready in all respects to meet the enemy'. We hoisted our battle ensign at the gaff, test-fired our guns, checked our ammunition loads, stacked our ready-use ammunition deck lockers to the brim, cut sandwiches, filled our Thermos flasks and positioned our tin hats and medical kits for immediate use. All these preparations were necessary to ensure the maximum fighting efficiency of the ship, but they also served another important purpose. By keeping everyone fully occupied in this way any tendency towards a build-up of tension among officers and men would be counteracted.

By midday we were only about thirty miles from Salerno and it was time to bring the ship to the second degree of readiness – semi-action stations. I had purposely held back the rum issue until this moment and it was now distributed, together with sandwiches and hot drinks. I gave each engine in turn a few minutes of maximum power, to blow it through and to check that it could be relied upon to meet any high speed emergency demands. Finally I went the rounds of the ship, having a cheery word with everyone and making encouraging remarks. It was at times like this that we resembled a large family, with myself as the father figure, a role I took very seriously, knowing how much all these young men relied on my actions and judgment.

We soon sighted some aircraft, apparently enemy ones, and puffs of anti-aircraft fire. We saw gunflashes from the hills behind Salerno and the smoke and explosions of shells landing there; and then the heartening spectacle of the ships that were firing them. As we neared the beaches the activity on land, sea and in the air progressively increased. There were ships of every type, ranging from Coastal Force and landing craft to cruisers and aircraft carriers, and from minesweepers and hospital ships to destroyers and monitors. Shore-based artillery was firing on the landing beaches and into nearby support craft, and our ships were shovelling out devastating fire into the enemy positions. Intermittently the Luftwaffe made bombing forays on our fleet and the Allied air forces attacked the German lines. Occasionally there was an aerial dog-fight. Everyone, both friend and foe, was feverishly engaged in his own particular task, seemingly without regard to all the other intense activity going on around. At times the noise was almost ear-shattering, with the thunderous roar of heavy naval

Salerno: Allied warships bombarding German positions, September 1943.

guns and the thudding of land artillery, the explosions of shells and bombs and the infernal row of aircraft engines. All this unpleasantness was in the most discordant contrast with the great natural beauty of the setting – a calm blue sea, a clear blue sky and the magnificent backdrop of one of the loveliest stretches of coast on earth.

In the mass of ships and the general confusion, it took some time to locate HMS *Hilary*. I found Commodore Oliver on the upper deck. He was hopping mad and brandished a naval signal form which he handed to me to read. It was worded in the strongest terms and stated that unless Montgomery could be spurred to get a move on there was a great danger of the whole Salerno operation having to be abandoned. The best hope of helping the army would be further reinforcements of big-gun ships. In a few short but picturesque sentences the commodore explained to me how desperate the situation had become. The Eighth Army, which had crossed from Messina into Italy on 2 September, had failed to make the planned junction with our Salerno forces and was still pinned down many miles to the south. Worse still, Montgomery had failed to secure his first main objective, the military airfield of Montecorvino, which was essential to bring our fighter aircraft within range of

Salerno to provide us with adequate air cover. Meanwhile Rear-Admiral Sir Philip Vian was doing the best he could with a force of five escort carriers. There was no problem in catapulting off the carrier fighters, but their comparatively low speed, the short runways and the complete lack of any wind combined to make return landings somewhat hazardous, with resulting loss and damage to the aircraft. As if all these adverse factors were not enough to contend with, there was a dangerous new menace – the radio-controlled glider bomb which was launched from aircraft. The previous day the American cruiser *Savannah* had received a direct hit from a glider bomb which had blown a hole in her bottom and had put her out of action.

Commodore Oliver instructed me to operate under American command in the western sector, where my immediate boss would be Captain Andrews in the USS *Knight*, a modern destroyer. Apparently the *Knight* had a roving commission, and if I failed to find her I was to use my own initiative and judgment in making MGB 643 generally useful, particularly in undertaking rescue work and firing on low-flying enemy aircraft. On no account was I to evacuate anyone from the landing beaches. 'We've all got to stay here and fight it out whether we like it or not,' were the commodore's last words to me as he turned to deal once again with the communications that kept pouring into his headquarters ship. What a man he was and what a heavy responsibility he bore.

Everything had been dealt with so quickly and efficiently that I was back aboard 643 within ten minutes of leaving her. We cast off and headed for the western sector. As we manoeuvred through the maze of ships I particularly noted the continuous patrolling of destroyers on the lookout for U-boats. War at sea is fully three-dimensional – mines and torpedoes under the water, gunfire on the surface and air attack from above. We cruised around but found no trace of the *Knight*. During the afternoon the Luftwaffe made sporadic attacks on our ships and there were several heavy Allied high-level bombing attacks on the German positions. But air raids can only be intermittent and are seldom very accurate. On the other hand the pounding of enemy targets by our warships was continuous and precise. In that calm sea they were rock-steady gun platforms and their targets were either stationary or slow moving. The results, which we could see through our binoculars, were devastating. What impressed us most were the salvoes from the 15-inch guns of the British monitors. The most intriguing noise from these was not so much the thunder of the discharge or the terrific crump of the explosion but the actual passage through the air of the huge shells, each of them weighing a ton or more. They emitted a surprisingly loud screaming whine – like a tram taking a sharp curve too fast.

The next day, 13 September, proved to be crucial, and it started in an ugly way. Just before dawn, the hospital ship *Newfoundland* was wantonly attacked by the Luftwaffe and sunk with dreadful loss of life. She was well away from the beach-heads and was brilliantly floodlit to display the huge Red Cross markings on her hull and decks. The battle of Salerno continued to rage throughout the day. During the previous night the Germans had launched mass attacks and had driven our forces still further back; they were now within two or three miles of our original landing points. With the coming of daylight, however, our big naval guns wreaked such terrible destruction on their armoured columns that their progress was halted; and there were exchanges of gunfire ashore as well as bombing raids by both sides.

During a lull in the early afternoon we went alongside the British Fiji class cruiser HMS *Uganda* in the hope of scrounging some bread and fresh water which we urgently needed; but they had helped other ships earlier and had no supplies to spare. A chief petty officer commiserated with our coxswain about the shortage of water as we cast off. We hadn't got very far when there was a tremendous explosion. The *Uganda* had been hit by a glider bomb. As she started to settle by the stern we raced back to see whether we could be of help. There was no panic and everyone aboard the cruiser went calmly about his duties. The coxswain spied his chief petty officer friend on the quarterdeck and shouted, 'Well, you've got plenty of water now, mate, haven't you?' Even at times like these lower deck humour was irrepressible. Damage control in the *Uganda* proved to be highly efficient and emergency repairs were soon carried out. We stood by for some time but were not needed. We heard later that although the *Uganda* was extensively damaged she managed to limp to Malta for dockyard repairs.

We resumed our self-imposed duty of patrolling along the coast. I decided to continue until we reached the extreme westward end of the landing areas. An attractive little inlet some four miles inshore of our patrol line caught my eye. I pointed it out to the first lieutenant and said, 'We're heading in there to have a recce. Pass the word around to guns' crews and engine room and tell them to be ready for instant action.'

We had just entered the inlet when all of a sudden a large number of US Rangers came running round a bend in the road about a mile away, obviously urgently seeking suitable cover. Ahead of them sprinted a naval officer, who threw himself into a dinghy and rowed towards us as though his very life depended upon it. Immediately afterwards a column of German Tiger tanks came rumbling and clanking round the bend. We raced to meet the dinghy, hauled the officer on board, made smoke and sped seawards. There could be no question but that in this case discretion was the better part of valour. Our small guns could hardly make a dent in the heavy armour-plate of the Tigers, but their powerful high-velocity guns could blow us out of the water.

When we were a safe distance offshore I paused to interrogate our unexpected guest, an Italian lieutenant-commander. He claimed that he had been acting as a liaison officer with the Rangers, assisting them with his local knowledge and connections. He was quite young and was evidently resilient. With a smile he thanked me for his rescue and said, 'It's much more pleasant at Capri. Why don't we go round there?' He added, 'It has an excellent MTB base.' I thought he must be joking. He went on, 'I can assure you that the Germans pulled out of there some days ago.' He finally convinced me that it was worth going to have a look. We desperately needed water, food and above all sleep, and Capri was only fifteen miles away.

We passed close in under Amalfi and as we brought Sorrento abeam we turned hard to port, heading for the breakwater harbour of Capri. We were tensely alert as we entered and sighted several Italian MTBs lying alongside the quay, but they offered no resistance. When we drew close enough our Italian lieutenant-commander had a parley with the captains and then explained, 'Everything is OK. They knew you were British. There are no Germans on the island and they're really glad to see us.' We berthed at the outer end of the quay, bows pointing to seaward and with our mooring lines 'singled-up' ready to slip at a moment's notice – I wasn't taking any chances. Having set watches we took turns to get some sleep, of which we had had precious little since leaving Messina.

Things happened thick and fast next morning. At about 8am a uniformed chauffeur brought down a splendid Alfa-Romeo for my use – 'with the compliments of the marchesa'. When I drove up the hill to the small town, people clapped and some threw flowers into the car. I began to

Capri from the air.

feel I was playing a part in a comic opera – as the Great Liberator. Everything seemed unreal. Here we were on this most beautiful of islands in perfect holiday weather, surrounded by friendly natives and away from the sights and sounds of battle. I made my first shore-going sortie a short one and was back aboard 643 by 10am. Just before 11 who should come steaming in but Captain Andrews in USS *Knight* in company with three 'D' MGBs. This was a welcome sight and I clambered over the rail of the destroyer as soon as she was tied up. I took an instant liking to Captain Andrews, who greeted me with, 'Who in the hell are you and what in the hell are you doing here?' He laughed when I replied, 'That's a good question, sir,' and proceeded to explain. With him were two well-known figures. I recognised one of them although he was in the unfamiliar 'costume' of lieutenant-commander USN; I noted that he was wearing the ribbon of the Distinguished Service Cross. It was the actor Douglas Fairbanks Jnr. The other man was the writer John Steinbeck. Fairbanks was executive officer (first lieutenant) of the *Knight* and Steinbeck was serving as a special war correspondent.

Captain Andrews was a man of action and got down to business at once. 'Get the skippers of those MGBs aboard for a meeting right now, Duggie boy.' Fairbanks hurried off to fetch them. While we were waiting, Andrews filled me in. I was interested to learn that he and the three MGBs had accepted the surrender of Capri two days before while on their way to the remote little island of Ventotene where they had hoped to capture Mussolini – but they were too late. He had been taken off just a few hours before.

When Duggie returned with the COs of the other three MGBs, I found that they belonged to the newly arrived 20th MGB Flotilla and that they were all British boats and crews though some of the COs were Canadian. The conference that followed was informal to a degree and produced some surprises. It was my first experience of being under immediate American command. First, Captain Andrews announced, 'You boys are now going to work under a different set-up. With 643 joining us we'll have a fleet of four boats, which will be divided into separate units of two boats each. So we will have a "dooty" unit and a "non-dooty" unit, which will change over every twenty-four hours at twelve noon.' The idea of having a clear twenty-four hours in harbour free of duty every second day sounded attractive. We had never had it as good as that before. Andrews continued, 'You've all got a nice little job tonight – we're going to capture Procida. So the duty set-up for tonight will provide two boats, 643 and 657, and the non-duty set-up will provide the landing party. Duggie and I will be close by in *Knight* to give you some back-up in case of trouble. You'd better start to get the ships and landing party ready. We sail at 20.00.' As we got up to go he turned to me, 'You lead tonight, Geoff, in that 643 of yours.'

Promptly at 8pm on 14 September our two crowded MGBs cast off and set course for Procida. The landing parties nearly doubled our usual complements. With me on the bridge was John Steinbeck, who had with him an amazing new invention – a wire-recording machine, one of the very few in the world at that time. Steinbeck was a most interesting and amusing character and it seemed no time before we were at Procida. We went straight in alongside the quay to enable the landing parties to jump ashore instantly. We then pulled clear so that we could manoeuvre to give covering fire if necessary. Nothing happened. As at Capri, instead of resistance, there was welcome, and instead of fighting there was embracing. The prepared instrument of surrender was formally and gladly signed by the mayor and the commandant of the small Italian garrison. Our 'invasion force' did strike one real problem – the difficulty of shaking off our enthusiastic and generous Italian hosts. We were soon safely back at Capri in what must have been the world's most beautiful Coastal Force base. We were still chuckling when we arrived there. It had been sheer comic opera again.

To me, the most entertaining part had been listening to John Steinbeck giving a blow by blow account on his wire-recorder as we rushed up to the quay at Procida and disgorged the landing parties. It made me feel that I was watching a film of the event rather than taking part in it. Equally unreal, in a different sort of way, was the account published in the *Sunday Express,* which I read some weeks later. It covered nearly the whole front page, and great banner headlines proclaimed:

GERMANS OUTFLANKED . . . AMERICAN NAVAL FORCE CAPTURES KEY ISLAND OF PROCIDA.

Whatever would our journalists dream up next?

The following night a similar and equally successful expedition liberated Ischia, the largest of the Neapolitan islands. After these three operations which we had all enjoyed so much there was something of an anti-climax. We carried out several patrols up the coast as far north as Anzio but never once encountered any enemy activity. The 'duty' and 'non-duty' arrangements worked like a charm, especially the non-duty part. The weather remained absolutely perfect and there were no air raids. Even the dominating mountain of Vesuvius, from which only a slim whiff of steam rose lazily in the still air, had a beneficent mien.

I did all the usual sightseeing on Capri, including visiting the remarkable Blue Grotto, but what intrigued me most were the residences of three notable people – Count Ciano, Mussolini's son-in-law; Gracie Fields, the singer and actress; and Axel Munthe, the Swedish doctor whose autobiography, *The Story of San Michele*, caused a sensation when published in 1929 and became one of the greatest best sellers of all time, being translated into twenty-five languages. We conferred with Captain Andrews and decided to respect Gracie's privacy and keep her delightful coastal villa securely closed. By a remarkable coincidence Axel Munthe's heir, a major in the Gordon Highlanders, was fighting at Salerno at the time and told his colonel about his house at Capri. Arrangements were made by us to send an MGB to Salerno to bring Major Munthe on a brief visit to the island so that he could assure himself that his treasures were intact. We didn't intrude on him and kept the property sealed thereafter. 'San Michele' was built largely from the ruins of the palace of the Roman emperor, Tiberius, and on the very site of the splendour of two thousand years ago.

We had no compunction about opening up Count Ciano's villa and having a thorough look at it. I took over the place myself for a couple of days and slept in a large bed in which Count Ciano and Edda Mussolini must have spent some happy hours. It was an extensive building and although ultra-modern and ultra-luxurious everything was in perfect taste – nothing jarred. The master bedroom had an unusual feature. Behind the bedhead were a number of press buttons giving remote control of all the window curtains and venetian blinds. When they were open, the view across the glorious bay to the sprawling city of Naples was nothing short of magnificent.

After ten halcyon days I received orders to sail for Malta. Never had we been so sad to leave anywhere as we were to leave Capri and our kind friends there, many of whom gave us some delightful parting gifts and souvenirs. We arrived in Malta early on the morning of 29 September, berthing near the destroyer anchorage in Sliema creek. A naval messenger awaited us. The envelope he handed me contained an invitation from an old friend to lunch with him that day aboard HMS *Nelson*, one of the greatest battleships afloat. In due course I walked over the hill to Valetta and Grand Harbour, where I stepped into one of the *Nelson's* picket boats, which was full of very senior officers. As I boarded the *Nelson* there was an interesting example of naval tradition and the emphasis placed on command rather than rank. Being by far the most junior officer in the picket boat, in accordance with naval etiquette I was last up the quarterdeck ladder; but I was the only one to receive a full ceremonial pipe. None of the others was so entitled. They were all staff officers and not in command of their own ships.

My host was on the quarterdeck to greet me; it was he who had advised the officer of the watch of my entitlement to be piped aboard. I must admit that throughout my service in the navy it always gave me a thrill of pride whenever this courtesy was paid to me. The *Nelson* seemed vast after 643, and indeed there could hardly have been a greater contrast. Her 16-inch guns could hurl projectiles weighing well over a ton apiece for a distance of more than twenty miles. MGB 643's heaviest gun fired a shell weighing six pounds over a very short range. The *Nelson* carried a complement of close on two thousand officers and men, against fewer than forty in 643. I had little time to look around the upper deck as I was hurried below with almost indecent haste. While we were sipping the first of several pre-luncheon drinks, an order was piped through the ship's loudspeaker system to the effect that no unauthorised person could remain on the upper deck and that we were to stay below until otherwise instructed. This was followed by the order to 'darken ship'. All blackout shades were drawn and deadlights clamped over the scuttles. Our incarceration lasted for two hours or more but we passed the time

HMS *Nelson*, 30 September 1943.
Left to right: Lord Gort, Air Marshal Tedder, Marshal Badoglio, Lieutenant-General Mason-MacFarlane, General Eisenhower, General Alexander.

pleasantly enough by swapping yarns, although I for one kept wondering what had sparked off such an extraordinary series of orders.

Just before I returned to 643 I was told that while we had been battened down below the formal Instrument of Surrender of Italy had been signed. Attendant ceremonies had been held on deck to mark this greatest achievement of the war against the Axis powers to date. The gathering had been a distinguished one. The leading participants were Marshal Badoglio for Italy; General Eisenhower for the United States; the three British Mediterranean Chiefs of Staff, Admiral Cunningham, General Alexander and Air Chief Marshal Tedder; the American diplomat Mr Murphy (President Roosevelt's personal representative in French North Africa); his British counterpart Mr Macmillan; and Field Marshal Lord Gort (Governor of Malta).

I thought that it was especially fitting that the venue for this great occasion should be a ship of the Royal Navy, which had achieved and suffered so much in the Mediterranean theatre. There was no doubt in my mind that the greatest of the Mediterranean Chiefs of Staff was Andrew Cunningham. We were all delighted when it was announced only a few days later that he had been appointed to the supreme post in the Senior Service – First Sea Lord and Chief of Naval Staff.

10

When I got back to 643 after my visit to the *Nelson* there was a message from Alan McIlwraith asking me to 'rejoin the flotilla'. I found that MGBs 645, 646 and 647 were all berthed nearby. Basil Bourne was still in command of 645. Tufty Forbes had been drafted to an appointment in the United Kingdom after the Sicilian campaign and his first lieutenant, Knight-Lacklan, had inherited 646. A manager of one of the British Home Stores branches before the war, Knight-Lacklan was well built, tough-looking and had a fair complexion. His most remarkable feature was his smooth 'india rubber' face which, like a poker player's, never seemed to change expression. Mountstephens's ship, 647, had rarely if ever been serviceable enough for operations since arriving in the Mediterranean. Before the war 'Mounters', as he was always called, had been a noted racing yachtsman and had won the Thames Estuary 'One Design' Championship year after year. Tallish and sandy-haired, he somehow contrived to be inconspicuous.

Our flotilla leader, Alan McIlwraith, who was quartered in 645, wanted to know all about our Salerno operations, some of which made him chuckle. He thought we might soon be needed in the Eastern Mediterranean and had made arrangements for 643 to be completely checked over and refitted as a matter of urgency, starting the next morning. Then he said, 'I've got some good news for you, Geoff. Your ship has been allocated a Mention in Dispatches for your work in Sicily. It's for you to decide who is to get it.' I had no hesitation in nominating our chief motor mechanic, Hunter, for this honour, and he was absolutely delighted when I told him about it.

Three or four days later Mac's prognostication proved correct. We had barely finished refitting, refuelling, reammunitioning and revictualling, when we were ordered to sail for Alexandria. This great British naval base in Egypt was twelve hundred miles away, the same distance as Gibraltar is from Southampton. It was a pleasant cruise in peaceful waters the whole way and our two stopovers for refuelling – Benghazi and Tobruk – could not have been more conveniently placed. An agreeable surprise awaited us at Benghazi, where we were greeted by our old friend Bobby Craven, who was in commmand of the small naval base staff. While we were there we lacked for nothing that he could provide. He was obviously very popular with the army and he explained how this had come about. 'You know how hopeless the dear old pongos are at making themselves comfortable, don't you, Geoff. They're still living quite crudely in tents but at least they've got electric light now instead of hurricane lamps – I rigged it all up for them. Come and see my power station.' It must have been unique and was certainly well worth inspecting, being put together mainly from the diesel engines and electric generators Bobby had salved from an abandoned Italian submarine. I was most impressed by his ingenuity.

At Tobruk we were greeted by gunfire from the shore battery. We had flashed the signal tower and had been granted permission to enter the harbour. We had formed up in line ahead with crews fallen in on deck – the usual naval ceremonial on such occasions – when, without even a warning shot, the army shore battery opened fire on us, fortunately with great inaccuracy. Frantic signals from our powerful lamps soon stopped this nonsense; not a man aboard had budged. This unfriendly reception was the more inexplicable as Tobruk was now far removed from any war zone. We had a few things to say when we went ashore and we did not tarry, leaving early the following morning for Alexandria.

We were given the best berths in Alex – at the jetty of the Royal Egyptian Yacht Club at Ras-el-Tin. This was quite close to the home of the notorious profligate, King Farouk, who had tried to betray the British cause to the Germans and had become so untrustworthy and recalcitrant that he was now a virtual prisoner in his own palace, which was ringed by British tanks. The yacht club was a centre of great wealth and luxury with a superb restaurant and other facilities. All these splendid amenities were made available to us but 643 and 646 were given little time to enjoy them.

Immediately on our arrival at Ras-el-Tin, Alan McIlwraith had hurried ashore to report to naval headquarters. He was gone for quite a time and returned with some very disquieting news. The situation in the Dodecanese was grave. Leros, Kos, Samos and other Italian-held islands had been occupied by British forces in September, but the Germans had managed to secure Rhodes, which was of key importance because of its airfields. The Luftwaffe, reinforced by squadrons from Italy and even from the hard-pressed Russian front, was present in the Aegean in considerable strength. In sad contrast, and in spite of the Allies' overall air superiority, only a handful of Spitfires had been provided by the RAF to help in the defence of the islands. Worse still, our one airfield – on Kos – had been lost when the island was recaptured by the Germans on 3 October. The only aircraft we would be likely to see would be German, as the Dodecanese were well out of fighter range from any British base. Throughout the war our main enemy in Coastal Forces had been the Luftwaffe, and it was depressing to find that after all this time they still seemed to hold the whip hand wherever we had to operate.

Once again, our sister ships, 645 and 647, had developed defects, leaving only 643 and 646 serviceable enough to proceed to the Dodecanese theatre. Mac transferred from 645 to my 643 and we were soon under way again on a four-hundred mile trip, this time to Beirut. Before leaving Alex we took on some extra food supplies. In the islands we would be on detached service for a long period, without any base and with no supply facilities. Water would have to be rationed. There would be no fresh provisions, not even bread. Fuel would have to be carefully conserved so we would not be able to use our electric cooker except on rare occasions. We would have to subsist on cold canned foods and hard tack.

We spent three fascinating days in the lovely city of Beirut. The headquarters of the British Army Camouflage School in the Middle East were centred there and the staff did a wonderful job in training and equipping us. We learnt that one of the tricks of concealment was to become a rock, just where a real rock was likely to be – off a promontory or among other rocks. By the time we left, our ships could be transformed into rocks in just three minutes flat. Our only hope of survival, with the Luftwaffe in undisputed command of the skies, was to operate at night and to hide up during daylight hours. Reports were coming through of ships enduring as many as twenty air attacks a day before being sunk or finishing with empty ammunition lockers and magazines. The story is told of a cruiser about to undergo yet another air attack. One of her anti-aircraft gunners was heard to remark as he blew up his life-jacket, 'This is the only fucking air support we'll get round here!' British army garrisons were still holding out with great courage in many of the islands and we all knew that, as it had done throughout the war, the Royal Navy would continue to support and supply them to the end, at any cost in ships and men.

On the third day at Beirut, we sent the two MGBS out with instructions to camouflage themselves anywhere on the coast within a fifteen-mile radius. Mac and I had arranged with the army for a spotter aircraft to check how effective our camouflage was from the air. Flying as low as 3,000 feet we had difficulty in finding our ships, although we knew what we were looking for.

That evening Mac, Knight-Lacklan and I were invited to

Kastellorizo harbour with the Turkish coast in the background.

the British Embassy for a dinner party, a veritable banquet which none of us really enjoyed. It rather reminded me of a funeral feast as our hosts were well aware that we were sailing for the Dodecanese in the morning. We cast off at first light, shaping our course for Paphos in Cyprus, where we stopped briefly for a final fuel top-up before heading for Kastellorizo, the southernmost island of the Dodecanese. This was another four-hundred-mile run – the fifth consecutive one – making a total steaming distance from Malta of two thousand miles, which seemed a fair way to come just to look for trouble in enemy waters. It was at Paphos, according to Greek mythology, that the goddess of love, Aphrodite, rose from the waves.

By 4am next morning MGBs 643 and 646 were feeling their way, very gingerly, through the many treacherous reefs off Kastellorizo. The darkness under the lee of the island was almost impenetrable and there were no navigational lights or marks to aid us; but somehow I managed to bring 643 into the little harbour unscathed. We glided up to a small ruined quay, where I went ashore on a cautious reconnaissance. I was thankful to find British and not German personnel. There were a handful of men belonging to a special combined army and navy unit acting as undercover caretakers. It was first light by the time they had shown me where our stores of 100 octane fuel in jerricans were buried in shallow dumps. Kastellorizo hadn't been invaded yet but was under constant enemy air surveillance during the day. They urged us to leave the island as quickly as possible, before we, like the town, were obliterated by wave after wave of dive-bombing attacks. 'That's where you want to go,' they said, pointing to a great fjord on the Turkish mainland three or four miles away. 'Don't come back here except at night, and then only to pick up fuel. If the Jerries start to land, we'll signal you by hanging a carpet on the wall of that house over there so you'll know to keep clear.' I thanked them and hurried back to 643. It was now daylight and I was appalled at the wanton destruction of the town. The scattered remnants of furniture, domestic treasures, ikons and children's toys shocked and enraged me.

We were only too glad to get away from the ghastly ruins of Kastellorizo and sail into the beautiful forest-clad fjord of Vathi. We had no compunction or fears about entering Turkish waters. Turkey at the time was thought to be on the brink of joining the Allies. Locked away in my cabin was a sackful of bribe money which had been given to us at the embassy in Beirut 'to secure the cooperation of Turkish officials when necessary'. We carried on up to the head of the fjord without seeing any sign of life or habitation. But we did spot a small stream tumbling down the rocks to a patch of beach. Any supply of fresh water was of the utmost importance to us, so we lowered the dinghy to investigate. All was peaceful, and the silence was broken only by the sound of our oars and the splashing of the cascade. Suddenly, when we were fifty yards from the beach, a noisy fusillade of rifle fire rang out. With high-velocity bullets buzzing around us like angry bees, we didn't wait to argue with our invisible attackers but returned to the ship with all speed. That was the only time we ever tried to land in Turkey.

We headed seaward again and four miles on found an ideal anchorage, protected from wind and weather on practically all sides. Complete silence reigned again and we turned in, watch by watch, to sleep. As soon as darkness

MGB 643 under camouflage netting in the Aegean, 1943.

fell we returned to Kastellorizo to fuel, and then patrolled the coast of Rhodes. We saw nothing that night nor on the many succeeding nights that we carried out various patrols round the islands of the southern Dodecanese.

Our camouflage nets worked like a charm. From inside we could see through them pretty well, and it was difficult to realise that from outside, at any distance, we were virtually invisible. In fact we felt quite exposed when we saw and heard German aircraft prowling around, sometimes at low level, searching for worthwhile prey like ourselves. As we were severely rationed for water, rarely had any cooked food or hot drinks and were confined to our ship day in and day out, we found that life became progressively more intolerable. There was nothing to look forward to and tempers became very frayed. It became increasingly difficult to maintain discipline and good spirit, vital ingredients in any recipe for survival.

After a while we gave up using our camouflage nets when sheltering in Turkish waters, on the grounds that the Luftwaffe wouldn't risk provoking the Turks by attacking us there. The removal of the nets lifted our dreadful feeling of claustrophobia and helped more than anything else to restore morale.

Occasionally, in an attempt to obtain fresh food, we tried 'fishing' with 2lb depth charges, which had been developed to deal with frogmen. We had no success in killing anything bigger than a sprat until the unexpected and welcome arrival of a British-owned supply caique, flying the Turkish flag and manned by Greeks, bringing petrol, water and ammunition for us. The captain showed me his papers and I noticed that he had recently been awarded the MBE for his services. He had originally been a sponge diver. We told him about our depth-charge fishing and he explained that any big fish in those waters always sank to the bottom after being killed or stunned by explosives. He retrieved the catch from the sea bed sixty feet below us and while he was with us we had some good fishing. All hands enjoyed two meals of succulent fish and we had no hesitation in starting up the generators to cook them. The Greek captain also brought up a few sponges. His visit was an absolute tonic.

There were all sorts of 'firms' working in the Dodecanese, notable among which were two élite formations – the Long Range Desert Group, whose speciality was causing dismay and confusion behind enemy lines, and the Special Boat Section, which was operating many caiques for supply and raiding purposes. SBS men were also frequently landed, sometimes by submarine, on German-held territory to gather intelligence and carry out sabotage. They put paid to large numbers of enemy aircraft as well as fuel storage tanks and bomb dumps. The various firms went quietly about their own business and little or no information was passed from one to the other, as a safeguard against any leakage under tough interrogation if a member should be captured.

It was probably for the same reason that we had not

been told much about what was going on. But on the evening of 16 November we received an urgent wireless signal which changed all that. Leros had surrendered to the German invasion forces and we were to proceed to the area forthwith. We realised this must signify the beginning of the end of these futile Dodecanese operations. Leros was a well-fortified island and the last worthwhile British outpost. The orders had been to hold it at almost any price; as we were soon to find out, the garrison had fought stubbornly and had nearly driven the Germans back into the sea. The main cost had been to the navy. Towards the end, most of the garrison's supplies had been transported to Leros by our submarines. They had even managed to bring in field guns, which were lashed to their decks with wire rope.[1]

The message about the fate of Leros galvanised us into activity. We didn't wait to prepare for sea but upped anchor and headed for Mandalya Gulf, the nearest suitable Turkish inlet to Leros. This was a hundred and eighty mile run by the rather tortuous route we decided upon. We were very much on the alert as we sped through the night, passing west of the island of Simi and east of Kos. We thought we might have to run the gauntlet of German guns and searchlights as we skirted Kos, and we hugged the Turkish coast, navigating 643 through rock- and reef-infested waters. For a short time we were caught in a searchlight beam but no guns barked at us. We were soon out of range and raced onwards in open water east of Kalimnos and Leros before turning to starboard into Man-

[1] Author's note. What went wrong with the Dodecanese campaign? Plans for the capture of Rhodes and other Italian-held islands in the Aegean (Operation 'Accolade') had been prepared by Middle East Command before the surrender of Italy, in the hope, among other political and strategic benefits, of bringing Turkey into the war. The Americans, however, did not share Churchill's enthusiasm for 'Accolade' and refused to support it; Eisenhower feared that such a diversion could tax Allied resources and delay the invasion of Europe in 1944. It has been suggested that the weakening of German arms elsewhere in Europe was enough to justify the Dodecanese venture, despite its failure. Even if true, this was cold comfort for those who fought there and in no way excuses the mistakes that were made both in the planning and in the execution of 'Accolade'. A few thousand troops and a few RAF fighter squadrons, which could surely have been spared from the Italian front, might well have ensured the success of the operation. At least, that's how it seemed to us.

dalya Gulf, which encompasses several sheltered bays. Dawn was breaking as we sighted a number of our ships tucked into one of these natural harbours.

We were soon making a round of visits among the MTBS and MLS assembled there and gleaned a lot of information about the desperate fighting, both ashore and afloat, during the past four days and nights. They had succeeded in sinking several invasion craft and escorts. One of the ships I boarded was ML 456, a 'B' type motor launch similar to my late beloved 339. The CO was Lieutenant-Commander Monckton, whom I hadn't seen since Great Yarmouth days when he was attached to Mickey Thorpe's flotilla and commanded a 'C' type MGB in the rank of lieutenant. 'Monkey' and I had been good friends and I had once made an operational trip in his MGB when he had been short of an officer. A small lightweight of a man with a pronounced stoop, he had been a professional Merchant Navy officer before the war. He was now Senior Officer of the 24th ML Flotilla. His ship was in a shambles after being hit by a shell off Leros during the night. The six or seven sacks on deck were a gruesome sight. They contained what was left of a number of his men who had been blown to pieces by the explosion.

We patrolled close inshore off Leros that night in the hope of finding some of the garrison we could take off, but without success. On the way back we had a lucky escape. In bright moonlight, as we ploughed our way through the gleaming quicksilver sea in close-up line ahead formation, one of the large radio-controlled glider bombs momentarily cast its sinister shadow before plunging into the water between 643 and 646. The violence of the detonation shook both our ships, doused us with water and scattered bomb fragments all over our decks, but we roared on undamaged and no one was hurt. The pilot of the launching aircraft had mistaken us for a single larger vessel and must have thought he had scored a direct hit amidships.

Our Leros patrols continued, and over the next few nights our various ships managed to evacuate a number of soldiers from the garrison. But by far the most spectacular incident occurred about 9am three or four days after the fall of Leros, when a splendid Italian admiral's barge swept into our bay. It was brought in by one of our Canadian colleagues, Lieutenant Tom Fuller, who had been acting as naval liaison officer with the army in Leros after losing his MTB in the initial bombing raids on Kastellorizo. Tom, a man of outstanding courage and remarkable ability, had been with us at Bone, and his first ship, a 'D' type MGB, had blown up there when petrol had ignited as the engines were being started. Apparently the Italian admiral's barge had been anchored in a small cove, and for some days before the surrender of Leros Tom had been making her ready for sea. The first suitable opportunity for escape had arisen that morning. He slipped the mooring and raced out in broad daylight under the noses of the Germans, bringing a number of senior army officers with him. When Alistair MacLean wrote his adventure novel, *The Guns of Navarone*, he used this daring escape as a central theme.

Soon afterwards, and again in bright moonlight, 643 and 646 went north to reconnoitre the historic island of Samos. As we approached the main port, Vathi, we were greeted by a pall of acrid smoke and smouldering fires from the ruins of the bomb-blasted town. It was a pitiful sight, but the story which emerged in broken English from some of the survivors horrified us even more. The Luftwaffe had attacked in force the previous afternoon and had played cat and mouse with the terrified and defenceless inhabitants, sometimes letting go their bomb loads and sometimes screaming down in power dives and dropping nothing. When the panic-stricken people tried to escape up the mountain road leading out of the town, the Luftwaffe came in to slaughter men, women and children with their cannon and machine guns. We were too late in leaving Samos to be able to reach our Turkish haven before daylight, so we pressed on at near maximum speed to reduce the time we would be vulnerable to attack from the Luftwaffe's bully boys.

It was nearing the end of November when we headed

south again for Kastellorizo. Off Kos we were picked up by searchlights, which almost blinded us at times as we tried to con our ship through a confusing maze of rocks, and it was perhaps only surprising that nothing worse befell us than hitting a submerged rock with our port outer propeller. The impact bent the shaft and made a mess of the propeller as well. We shut down that engine and carried on with the other three. A garrison had again been built up at Kastellorizo and we had been ordered there to assist in its evacuation. When we reported to Colonel Ruffer, the Administrative Officer on the island, we learned that two or three destroyers were making a dash to Kastellorizo that night, and that it was our job to ferry the five or six hundred troops out to them. There was no need for him to tell us that everything would have to be done with the greatest speed and efficiency, so that the destroyers could get as far away as possible from the Dodecanese and the Luftwaffe before daylight. Once we had completed our planning with the colonel we set to work to replenish our water and fuel supplies.

We had a further talk with the colonel in the afternoon. He was a regular army officer of the best type, and we found him a pleasant and businesslike man to deal with – except in one respect. He had asked us to destroy the NAAFI store on the quayside by gunfire. 'Let's have a look at it first,' I said, hoping we might find something worth taking away. We did. I pointed to a number of cases of Scotch whisky. 'May I take those on board, before we shell the place?' 'Not unless you pay for them.' 'But otherwise they will only be destroyed.' 'That's quite a different matter. If you take them away without paying, that would be a form of theft.' In spite of all my efforts at persuasion, he remained quite adamant on this point. I had very little money on me, but it was finally agreed that if I handed over a promissory note I could take the whisky. What rejoicing there would be when I shared this out with the other officers in the flotilla! We had been kept on pretty short rations in the Mediterranean, only being allowed to buy an issue of one bottle of spirits per month. I thought the colonel's attitude, honourable though it might be, was quite ludicrous under the circumstances. Then it dawned on me that in matters of personal integrity we were just as ridiculous. We had a very large sum of bribe money on board which did not have to be accounted for, and yet it would never have entered our heads to dip our hands in the sack – not even to buy the whisky that the colonel could have written off as having been destroyed.

It was hard work embarking, ferrying out and getting the troops aboard the destroyers in pitch darkness. We took fifty to a hundred at a time, and to maintain stability we had to pack a lot of them down below. Each man came on board with full equipment, his rifle slung from his shoulder. These were good troops – they belonged to the 4th Frontier Force Rifles, an Indian Army regiment – and they were well disciplined. They had been taught never to part with their weapons, but we had to separate each soldier from his rifle to squeeze them down our hatchways. They resisted strongly at first, until we hit on the idea of holding the rifle momentarily and passing it back once the owner was below.

By one o'clock in the morning we had completed our exhausting chore and set course direct for Alexandria, almost due south and easily three hundred and fifty miles away. Our ships were in a shocking mess after transporting the troops, who couldn't help inflicting a lot of damage with their great hobnailed boots. The woodwork was bashed and scored from rifle butts. We were still uncomfortably crowded with passengers as the colonel had asked that he and many of his officers should be given passage to Alex with us. We also had on board an RAF group captain who had been acting as air liaison officer with the garrison. We reflected sourly that this must have been something of a sinecure as there were no British air forces in the Dodecanese to liaise with.

The long trip to Alex soon became an ordeal. We motored on for the rest of that night, all the next day and all the following night. Our ship's officers had given up their bunks to the guests so there was nowhere for us to lay our weary heads. We fed our passengers as best we could from our meagre rations and, ironically, plied them with the very liquor they had refused to give us – none of us ever drank at sea. We curved into Alexandria Harbour at 7am on 29 November – dead on our ETA – and headed joyfully for the Royal Egyptian Yacht Club jetty. Holway nudged me and pointed to starboard. Very close abeam and keeping station on us, with brass funnel gleaming, was the steam pinnace of the Commander-in-Chief, Mediterranean; and standing on deck and actually saluting us was Admiral John Cunningham in person. What we had done to earn such an unusual tribute, I don't know, but it was none the less inspiring for that. Before we could gather ourselves sufficiently to sound a pipe as a salute, the pinnace had sheered off as silently and as unobtrusively as it had arrived.

We berthed at Ras-el-Tin and disembarked our passengers. We were nearly asleep on our feet. The colonel and his officers were generous in their thanks and invited us up to their mess for dinner – a delightful occasion that we enjoyed a few days later. The group captain, however, left us with barely a word. No one was sorry to see him go.

11

Immediately on our return to Alexandria, Basil Bourne and Mountstephens, with Mac as Senior Officer, set off with MGB 645 and MGB 647 on a special mission to the very area we had so recently left. When they came back a week later escorting a badly damaged Greek destroyer, the *Adrias*, they were greeted with congratulatory blasts on ships' sirens from all around the harbour. They had pulled off a most unusual and daring feat of salvage in enemy waters. The *Adrias* had been beached after losing most of her bows on a mine. The Greek survivors and the crews of 645 and 647 carried out make-do repairs, using tree trunks they felled in the nearby forest. They jettisoned enough fuel and ammunition to enable the destroyer to be floated off, and managed to bring her back to the safety and repair facilities of the naval dockyard at Alexandria.

Soon after their triumphant return we were intrigued to see an admiral with a considerable entourage advancing towards us along the yacht club jetty. He was on board before we could arrange any ceremonial reception. One of his staff asked my name and presented me. It was Prince Paul of Greece, who wished to invest me and the other COs who had salvaged the *Adrias* with an important Greek order of chivalry. I hastily explained that the only two ships concerned were the MGBs berthed alongside. The decoration was thought to be the Order of Chastity, Third Class. The ragging and mockery that McIlwraith, Bourne and Mountstephens had to endure for months afterwards must have nearly driven them up the wall, especially as the award was in fact the War Cross.

There was great jubilation when we shared out the excellent Scotch whisky which we had acquired in Kastellorizo. In Egypt there was a locally distilled whisky available which was indescribably foul in smell and taste and probably highly dangerous to drink. Occasionally touts would offer black market stuff at exorbitant prices; it was unlikely that the contents were true to label in this land of trickery and deceit. In Alexandria and Cairo there seemed to be a complete inversion of Western moral standards. Successful theft, vice and even murder were more to be admired than condemned. Thieves would walk straight up to you in daylight, grab your fountain pen or your watch and disappear into the crowd before you realised what had happened. No loaded truck was safe without an armed guard in the back, and I saw quite heavy goods removed from lorries when momentarily stopped by traffic lights right in the centre of the city.

There were so many ships needing extensive repairs after the Dodecanese operations that there was no possibility of docking 643 to have her propeller and shaft mended for some time. However, this suited us quite well because we were overdue for leave. We had a great time. One of our favourite ports of call was the fashionable Union Bar, where for the first time we saw the delightful notice 'No Air Commodore under 18 years of age will be served, unless accompanied by his mother,' painted in prominent lettering. I took the opportunity of visiting Cairo. Impressed and interested though I was by the pyramids and other antiquities, my greatest pleasure was staying in the splendid New Zealand Services Club there. The sound of New Zealand voices and the natural warmth and friendliness of my compatriots filled me with nostalgia for the homeland that I had left more than ten years before. I spent several days at the club and was thoroughly spoiled by everyone. This was mainly due, I think, to the fact that no one could remember a naval officer having been resident there before.

It was not until a couple of days before Christmas that the powers that be were able to get us into the floating dock. This was close to the submarine mother ship HMS *Maidstone*, which was ready to provide the officers and men of 643 with excellent accommodation in individual messes, first class meals and many other facilities. She had been specially designed and built for this. Everyone aboard the *Maidstone* went out of his way to be helpful and we were offered the Christmas dinner of our choice.

I was sure that our ship's company would be delighted, but I found that they had other ideas. They wanted to have Christmas dinner with their officers and to eat it aboard their own ship, deprived though she was of electric light and other services while in floating dock. With the cheerful cooperation of the *Maidstone* we collected our stuffed turkey and its trimmings from their galley and carried it in covered dishes to 643. Drinks were on the officers. Nearly everyone proposed a toast or made a speech and there was a wonderful spirit of camaraderie. It was one of the happiest Christmas Days I have ever spent.

The crew had also planned and had saved up for a New Year's Eve party at the Fleet Club, to which they invited all the officers. It was an equally memorable occasion, well organised to the last detail. Dinner had been ordered in advance and various crew members entertained us from

Officers and ratings of MGB 643 in the Mediterranean.

the club's stage. I was constantly plied with drinks and it was clear that the object was to put me under the table. It was not achieved, but as I weaved my way back to the ship I found that Chief Motor Mechanic Hunter never left my side. 'Coxswain's orders, sir,' he said. 'He appointed me to look after you and make sure you didn't get into any trouble or do anything that would disgrace our ship.'

Our lotus-eating leave in Egypt came to an end early in the New Year, when we moved out of the floating dock and made ready for our next assignment, which was to be another detached service job. This time the scene was to be Yugoslavia. Marshal Tito had been forced to withdraw to the island of Vis and concentrate the battle-scarred remnants of his partisan forces there for a last desperate stand against the German invaders. We had enjoyed our break in Alex, but we were starting to think that it was possible to have too much of a good thing and we were more than ready to return to active duties. On 10 January 1944 we set out on another long voyage of eighteen hundred miles – fantastic by comparison with Coastal Force operations in the North Sea and the Channel. There were only the two of us again – 643 and 646. Alan McIlwraith had been promoted and was now deputy to the new Captain Coastal Forces, Mediterranean. This was Captain Stevens,[1] who had previously commanded a destroyer flotilla. Basil Bourne had also been promoted and was now a lieutenant-commander, replacing McIlwraith as Senior Officer of our flotilla; but he, with Mountstephens, was still stuck in Alex, both their ships again being unserviceable. I had been made a qualified officer, a new distinction, which gave the comparatively few RNR and RNVR officers selected full equality and seniority with regular RN officers. I was to act as Senior Officer of our unit until such time as Basil Bourne could follow on to Yugoslavia.

From Alex to Malta we were on familiar territory, stopping over again at Tobruk and Benghazi; but from Malta onwards we forged our way into what was an entirely new part of the area for us, the Adriatic. Our first port of call was Brindisi, where we berthed alongside a remarkable new Coastal Force base, which floated and was mobile. It was the *Miraglia*, a large ex-Italian navy seaplane carrier and mother ship; had she been specially designed for it she couldn't have been better suited to her new role. She had plenty of good accommodation, well-equipped machine shops and repair bays, and excellent petrol storage and refuelling facilities. Furthermore, she had ample crew to cope with all our needs as well as her own. She made a pleasant contrast to our previous mother ship, HMS *Vienna*. The original Italian naval captain, officers and crew were retained, and they worked efficiently with our Commander

[1] Later Vice-Admiral Sir John Stevens (b. 1900). Commander-in-Chief, America and West Indies Station, and Deputy Supreme Allied Commander, Atlantic, 1953–1955.

Coastal Forces, Western Mediterranean, Commander Welman, and his base staff, who also lived on board.

I had met Commander Welman before, when he was in command of HMS *St Christopher*, the Coastal Force training establishment at Fort William, and I had a cordial relationship with him. As a young officer, Welman had served with distinction in coastal motor boats in the First World War and had been awarded the DSO and two DSCs. He had spent some time in Yugoslavia with the partisans and it was mainly due to his influence and good judgment that Tito came to concentrate his forces in Vis, which Welman had already decided would be the best operational base for Coastal Forces. He was therefore able to give me a thorough briefing on the Yugoslav situation, mentioning that 'the old man' (Winston Churchill) was now getting personally interested in it.

Captain Stevens was spending a few days aboard the *Miraglia* after a visit to Vis. He sent for me. 'I've been studying your records, Hobday,' he said, tapping a file in front of him, 'and I think you are overdue for some advancement. You've had nearly three years now of sustained front line service in Coastal Forces. I'd like to keep you with us in the Mediterranean and promote you to Senior Officer of a flotilla of your own. But in fairness I must mention an alternative appointment opportunity for you – a more responsible command, which could be a frigate or an escort destroyer. They're building a lot of new ones now.' I told him I could be interested in either possibility. 'I want you to think it over carefully. Come back and see me in the morning.' He shook hands and showed me to the door.

I was delighted with this turn of events and was like a cat on hot bricks until I saw Captain Stevens again. I had thought hard about the two possibilities and in the end plumped for a 'more responsible command'. I'd had a long and lucky innings with Coastal Forces but it could not last indefinitely. It had been my ambition for some time to command a 'real' warship. A bonus was that I would be sent back to England to take over my new ship and would be able to see my wife and son. Captain Stevens seemed happy about my decision and said he would forward a recommendation through the appropriate channels that very day. 'But,' he said, 'it will probably take four or five months before there is a ship available, so don't get restless. You would be wise not to mention our talk to anyone.' Some weeks later he let me know that the recommendation had been 'approved by their Lordships', but that did not mean that there would be any immediate action.

Our last stop en route to Vis was at Bari, which was then the main supply port for the Allied forces in the Adriatic. Only a few weeks before – on the night of 2 December 1943 – it had been the scene of one of the most successful raids made by the Luftwaffe during the Mediterranean war. They achieved this with only twelve bombers, which came in low from seaward just after dark and sank no fewer than seventeen large cargo ships, laden with vital military supplies. Many other ships were severely damaged and over a thousand casualties were inflicted in the port area. It was a real disaster for the Allies. Coastal Forces suffered badly too, and many of our craft, assembled in Bari for operations in Yugoslavia, were put out of action for several weeks. Perhaps even worse was the loss of sixteen Rolls Packard Merlin engines and a wide range of spares when the ship carrying them was blown up and sent to the bottom.

I reported at once to Captain Campbell, the Naval Officer in Command, for onward routing instructions, but we were held up for three days. There was a pleasant New Zealand Forces Club in the town, and there I bumped into several old friends from Auckland including Peggy Robertson, a very comely WAAC sergeant who had been a prominent member of our social set at home. Peggy was a real stunner with a fair complexion, blue-grey eyes and the figure of a small Venus. I fixed a date with Peggy but when we met at the big Hotel Imperiale she was refused admittance because she was not an officer. I was at a loss until Peggy suggested that we went to the sergeants' club nearby – where I was refused admittance because I was an officer.

MGB 646 in the Aegean, November 1944.

It then occurred to me that I had a perfectly good ship in the harbour with all the necessary amenities and, as I was in command, no one could gainsay us there.

Peggy was absolutely fascinated by our little ship. I rang for the coxswain and told him that we would be making a quick tour and might look into the mess decks briefly. He left with a huge grin on his face, and we heard him whistling 'Kiss me goodnight, Sergeant Major' as he went along the alleyway. He returned to say that the crew would appreciate it if the sergeant and I would be their guests in the seamen's mess deck. When we rose to leave after an enjoyable evening the coxswain started to whistle that tune again. Peggy laughed. 'All right, coxswain,' she said. 'Fall the boys in and I'll kiss the lot!' She did it properly too, giving every man a warm embrace, except the coxswain who, much to the delight of everyone else, only received a perfunctory peck.

Next morning I saw Captain Campbell again. He had a warning to give me. On no account was I to give passage to Admiral Sir Walter Cowan on my trip to Vis. Cowan was a full-blown admiral who had been retired well before the war. He had won the DSO in the Sudan serving under Lord Kitchener in the 1890s. He was now seventy-two years of age and was trying to cross to Yugoslavia to fight as a soldier with the partisans. The authorities didn't want the old gentleman to get himself killed. The NOIC said, 'He's as sharp as a razor and is bound to put two and two together and guess that your ship is heading for Yugoslavia.' He went on to explain that the admiral had just come out of the front line in Italy where he had been fighting with the New Zealand infantry at Monte Cassino. Before that he had been in Tobruk. Apparently, when the South African general, Klopper, had surrendered there he told him to his face that he was a coward and a disgrace to his uniform. He then made his way to the outer perimeter to meet the enemy single-handed, jumping on to the leading tank and firing his pistol through a slit. He was taken prisoner but later exchanged because of his age.[1]

When I returned to 643 I found a small, wiry man, with innocent blue eyes, who looked very fit and sun-tanned. Sure enough it was Sir Walter. He seemed charming – until I had to refuse his request for passage to Vis. Finding that neither cajolery nor threats would move me, he made the most blistering comments on my lack of guts and enterprise. The verbal horse-whipping I had to endure from this fiery little man made me feel quite sorry for that South African general. By the time Sir Walter had finished with him he must have heartily regretted his decision to surrender. As I saw him over the side, the admiral shook his fist at me. He was consumed with rage and frustration and shouted, 'You haven't seen the last of me, Hobday. Nothing is going to stop me from getting over to Vis. Mark my words.'

[1] Author's note. I now know that much of this story is incorrect.

Lieutenant-Commander MC Morgan Giles at Komiza, Vis, 1944.

Late in the afternoon we tied up to the stone quay of the medieval fishing town of Komiza on the island of Vis, and I reported to Lieutenant-Commander Morgan Giles,[1] who was in command of the Coastal Force base that was in process of being established there. Morgan Giles had been awarded the George Medal for rescue work among the blazing oil and exploding ammunition ships in Bari harbour after the disastrous German air raid. A tall, spare man, he always moved at high speed both mentally and physically. His family firm were noted yacht and power boat builders at Teignmouth in Devon.

Events were to prove that we couldn't have had a better man to back us up than this competent and enthusiastic officer. He quickly gave me a general appreciation of the

[1] Later Rear-Admiral Sir Morgan Morgan-Giles (b. 1914). MP for Winchester, 1964–1979.

situation. Coastal Forces had been operating on a limited scale in Yugoslavian waters for some months. The fast 70-foot Vosper MTBs had been making sporadic strikes on enemy shipping and a number of 'B' type MLs had been running arms and ammunition to Tito's partisans. Morgan Giles had participated in and directed this latter operation. Early on he had made a very advantageous and far reaching decision to supply the partisans with German equipment. He had no difficulty in procuring all he needed from the dumps of captured arms and ammunition in Italy; practically the only arms the partisans already possessed were those they had taken from the Germans.

As a result of Churchill's insistence, the scale of Coastal Force operations was to be increased as speedily as possible and 'D' type MGBs and MTBs were to play an important role. Three 'D's had arrived a few days earlier and the addition of my two brought the strength of the force up to five. I was to act as Senior Officer of the 'D's for the time being. Morgan Giles went on to tell me about the incredible bravery and hardiness of the partisans. The shootings of hostages and other atrocities had strengthened rather than undermined their resolve. Women were fighting as front line soldiers beside the men. However, although we were helping them the partisans could be very prickly to deal with.

Tito's forces had no artillery or tanks, and had to rely on rifles, hand grenades and a few machine guns. There was no aerodrome on the island, not even an airstrip. In the early months of 1944 there was a very real threat of Vis being invaded, and the main responsibility for defending the island lay on us. But attack is often the best means of defence. Our job was to harry and sink German ships, and eventually isolate the enemy garrisons in the Dalmatian islands by completely cutting their lifeline of supply and communication.

The day that followed was a busy one, with further discussion with Morgan Giles and conferences with Admiral Cerni, who was in charge of the partisan navy, and with the COs of the other four 'D' boats in my unit. In the evening we were requested to attend an entertainment at

Komiza by Harold Garland.
This oil painting, commissioned by the Coastal Forces Veterans Association, was presented to the Yugoslav ambassador in 1981. It now hangs in the war museum at Split.

the fish factory, where in normal times sardines and anchovies were canned. There I not only met partisan soldiers, both male and female, for the first time, but also a couple of smiling, blue-eyed Russian military advisers. At first I thought the girl soldiers were remarkably well 'stacked', until I realised they were carrying grenades in their front tunic pockets. The 'entertainment' consisted mainly of short, almost charade-like, political and patriotic plays which, although secretly amusing to us, were attended to with great concentration and greeted with much applause. As, presumably, a fairly large percentage of the partisans were illiterate, this seemed to be an ideal means of instilling Tito's official propaganda into their unsophisticated minds.

I led units on patrol nearly every night and we had some early successes. On the night of 3 February 1944 we intercepted a diesel-engined supply schooner. On the spur of the moment I decided to try to take her by boarding instead of just sinking her by gunfire. Our manoeuvre was completely successful in that there were no casualties on either side and we took fourteen prisoners. As the boarding party returned they grabbed a couple of cases from a pile on deck before I sank the schooner by depth charges. This was a mistake. I was full of chagrin when I learnt from the prisoners that the schooner had been laden with food supplies, including many delicacies destined for German officers. My mortification was complete when the cases seized from the deck cargo were found to contain only low grade ersatz bully beef.

Our next encounter was on the night of 8 February, when we sighted a tug stealing along the coast, heading for the German-held harbour at Rogoniza, which we always referred to as 'Roger's Knickers'. We had to deal with her quickly before she slipped into safety and had no option but to sink her with our guns, along with the pinnace she was towing. Tugs were of particular value to the enemy and the loss of this one would hurt.

It was only a few days after Sir Walter Cowan's angry departure from 643 in Bari that, to my amazement, he came on board in Vis. As before he was dressed in khaki shorts and shirt with no badges of rank. On this occasion he was positively jovial and shook my hand warmly. 'Well, here I am, Hobday, as promised.' He seemed in such a good mood that I took a risk. 'As threatened, sir!' He liked that and laughed.'Sorry about our little fracas in Bari. Let's forget all about it.' From that moment I knew I had a friend. When I asked him how he had managed to reach Vis he simply said, 'Oh! I just got a lift with some friends', and left it at that. We never did find out how he had made it.

The doughty little admiral soon became liked and respected by all of us, British and partisans alike. A few months later he distinguished himself in action when he won a second DSO – forty-six years after the first one – for his 'gallantry, determination and undaunted devotion to duty' as liaison officer with the 2nd Commando Brigade during fiercely contested raids on the islands of Solta, Mljet and Brac. He left Vis not long after I did but we continued to keep in touch with each other. In a letter I had from him after the award of his DSO he wrote, 'As to this great honour they have done one I feel quite undeserving of it and rather ashamed of the rubbish that was in the newspapers because as you all know I was only a sort of joy-riding onlooker and no value to the war effort – while all of you were never idle.' Another letter advised me that he had sent a case of wine from the cellars of his Warwickshire estate to my wife to await my return. Sir Walter was a great horseman and took more pride in being made Honorary Colonel of the 18th King Edward VII's Own Cavalry, the regiment he had served with in North Africa, than in anything else.

The atmosphere at Vis was exhilarating and quite the antithesis of the sense of hopelessness that had hung over us in that ghastly Dodecanese campaign. Here the odds were still stacked against us, but we were well led by Captain Stevens and Lieutenant-Commander Morgan Giles, and we couldn't help but be inspired by the daring and fervour of the partisans, who all fought like VCs. We were attacked in harbour by the Luftwaffe nearly every day – often several times a day – but kept our ships dispersed as much as possible and sustained surprisingly little damage. In fact we suffered more from storms than from air attacks, owing to that Adriatic weather phenomenon, the bora, which in winter and early spring howls down from the mountains with sudden and terrible ferocity and can gust up to seventy knots.

Although when on station at Vis we generally operated every night to seek out the enemy, we had frequent breaks back at our *Miraglia* base in Brindisi. This was a policy instituted by Captain Stevens which enabled the ships to be properly serviced and maintained, and kept the crews in good fettle too. We had never before worked so happily and efficiently. One morning in Brindisi towards the end of March I was puzzled when I came on deck to find the ship covered in nearly half an inch of greyish dust. Vesuvius, the great volcano outside Naples, had awakened from its fitful slumbering and had erupted. This volcanic dust had been blown from the other side of Italy. For us it was just a mess to clean up but for those near the angry mountain it was far worse, and the Allies did in fact have to write off many aircraft and motor vehicles as a result.

One of the good things about these visits to Italy was the freedom from air raids. The terrible cost of the Bari raid had led to a drastic tightening up of both our air force and anti-aircraft defences, which were now in a state of constant readiness and vigilance. The Luftwaffe soon found that any further surprise attacks of the Bari type were not on, and concentrated their activities more in the battle zone or on less dangerous targets like Vis.

So much happened in the next few weeks that it is difficult to recall all the events, let alone record them in chronological order. More and more MGBS, MTBS and MLS kept arriving in Vis and the number of Coastal Force craft operating from there soon exceeded anything I had known in other bases. Detachments of Royal Marine and army commandos were sent to the island and a group of US Army engineers set to work to build an airstrip on the difficult mountain terrain just behind Komiza. British Trowbridge class destroyers made a number of forays at night to shell the German invasion concentrations and when returning to Italy just after dawn would steam round the island close inshore. There were generally five or six of them and, as was intended, the sight of these splendidly belligerent ships gave a tremendous boost to the morale of us all. At times we broke into spontaneous waving and cheering as they passed by.

Another reassuring event was the arrival of an army medical officer, Major James Rickett, with a small team of orderlies. Previously we had to tend our wounded ourselves, without any professional aid. James Rickett was a fine man and a quite remarkable surgeon who accomplished much under almost impossible conditions at Vis. After the war a book was published about his adventures there.[1]

I must have been one of Rickett's first customers. I needed his surgeon's knife not for wounds suffered from enemy action but for a much lowlier purpose – piles! These had probably developed from standing for long hours on the bridge and were getting more and more painful. Rickett had been installed in an empty house on the top of a 1,500-foot hill that rose just behind Komiza Bay and I was driven up the twisting unfenced mountain road in a jeep. Fortunately the piles were external. Rickett cut them out then and there without using any anaesthetic, which was in such short supply that it was reserved for much more serious operations. I had to kneel on a table close to a window right on the village street, along which peasants were returning from their work in the fields. My bottom must have presented a most unusual and interesting sight to all who were fortunate enough to pass by at the time.

Next morning, after a fairly painful night out on patrol, I had a meeting with Admiral Cerni at Tito's headquarters. He told me that the Germans had put prices on the heads of a number of British officers, including my own. This sounded very melodramatic but the colonel's intelligence information was almost invariably right, so I took some heed. He then handed me a five-pointed red star about the size of a cap badge. On it was embroidered a yellow hammer and sickle. As I stared at it, he drew himself up and said with great dignity in broken English, 'That is the badge for a commissar. You show that to our people if you need help and they will do anything.' I knew that in being made an honorary commissar I had been accorded an unusual privilege and I still treasure that little red felt star to this day. Subsequently it became a standard practice to give similar commissar badges to leaders of flotillas operating from Vis.

Before I left the partisan headquarters Admiral Cerni

[1] *Island of Terrible Friends* by Bill Strutton (London, 1961)

asked me to give him a first-hand account of the attack we had made on Sibenik a few nights before. He was personally interested because his intelligence service had given us the information that German naval headquarters was now located there. Our incursion into the apparently impregnable Sibenik 'fjord' had not been planned. It was a purely spontaneous decision generated by the conjunction of two fortuitous circumstances – the strength of my patrol unit and the state of the moon at that particular time on that particular night. I had four good 'D' boats on that patrol, and as we made our way up the coast in bright moonlight it suddenly struck me that the eastern side of the long, narrow Sibenik inlet would be in the deep moon-shadow cast by the precipitous spine of hills which rose sheer out of the waters of the fjord. This was an opportunity not to be missed and I quickly outlined my plans to my consorts. Two boats would proceed to create a diversion by engaging the attention of the island-based shore battery that guarded the entrance to Sibenik, while 643 and the other 'D' boat would try to sneak unobserved into the harbour under the shadow of the cliffs. These hastily conceived plans were the best I could think of at the time. We managed to slip inside undetected, but the four-mile trip up the fjord was a real test of nerve. Both ships ran slow ahead on a single silenced engine to reduce wash and noise to the minimum. Hoping to achieve the greatest possible element of surprise, I had determined to make my attack from the inland rather than the seaward side of the base, which entailed going a mile past it before turning and making our run for the target and the open sea.

It was by far the most successful and exciting attack I had ever made. As soon as we turned we crash-started our other three engines. With open throttles and exhausts we each unleashed our full six thousand supercharged horsepower, which drove us down on to the German base in great clouds of smoke and spray while we poured out thousands of rounds of assorted armour-piercing, incendiary and tracer shells and bullets. The deafening noise of our engines and guns must have been fearsome indeed, bringing a nightmarish awakening to the enemy. The sheer audacity of our penetration of their seemingly secure stronghold and the direction of our attack appeared to have utterly confounded them. There was no return fire.

We received unexpected aid from the German guns guarding the entrance to the inlet. Our ricochets were bounding towards them and the German gunners must have assumed that the partisans were making a land attack on the base. They fired salvo after salvo into the hills behind it. This solved the problem of getting out past the battery without taking a real pasting. The enemy were still concentrating on shelling the wrong target when we roared through the entrance to the open sea and rejoined our unit, absolutely unmarked and with no casualties whatever. For the Germans it was an entirely different matter. We had created havoc at their base and many ships and buildings were blazing as we left. Our brief visit must have hit hard psychologically as well and they must have fairly smarted from their loss of face.

Morgan Giles found it difficult next morning to believe that we had done so well at Sibenik, but when all the details had been recounted he laughed until the tears rolled down his cheeks. The bit he liked best was the magnificent help we were inadvertently given by the German guard battery. He must have communicated the news to our boss in Italy, Admiral Morgan, who was Flag Officer, Taranto and Adriatic. The following morning a messenger brought me an official naval signal which read, 'Congratulations on singeing the German admiral's beard', thus making a very flattering analogy of our little affray with the tremendous destruction Sir Francis Drake brought down on the Spanish fleet at Cadiz with his fire-ships. Anyway, we were all frightfully chuffed by this signal from Admiral Morgan, which confirmed that the spirit and flair of the Royal Navy remained unchanged.

Overleaf: Tracer at night.

12

Thanks to the ever-increasing support we were now giving the partisans, they had gone on to the offensive and there was a considerable resurgence of their guerilla activities on the mainland, the emphasis being placed on disrupting the German supply routes. The partisans mounted so many road ambushes and rail demolitions that they forced the enemy to rely on sea transport for most of their needs. The Germans had to commandeer every sort of vessel they could lay their hands on, including diesel-powered coastal schooners, trading ships and small tankers to carry fuel and water, the more so because their increasing need for additional escort vessels compelled them to convert most of their landing craft into gunboats.

We decided to take a leaf from the partisans' book and forget about conventional methods of warfare. Because our guns were too small to sink larger ships and because we had commandos available, we came to make a bold and far-reaching decision. As in the past, we would take ships by boarding, but would deploy our firepower and our boarding parties in entirely new ways. Every 'D' boat going on patrol now carried a detachment of Royal Marine Commandos, armed with the Lanchester carbine (a deadly, compact sub-machine gun, ideal for fighting at close quarters), short bayonet and hand grenades.

In the preliminary softening-up stage of the attack, our 'D' boats converged on the enemy from different directions to split his fire power, while our guns were systematically used to destroy control and communications and to disable his engine or boiler room. Our forward 2-pounder pom-pom concentrated on the enemy's bridge, the 6-pounder on his engine room and the Oerlikon twin cannon, together with our many machine guns, were used to spray his decks. Meanwhile our commandos lay hidden and safe on the blind side of our ships until we crashed alongside and they stormed aboard the enemy from widely dispersed points. One party would take the bridge, another the engine room and a third would round up prisoners, forcing them on to the foredeck. Immediately the boarding parties were away, we would sheer off and stand by with our guns trained on the enemy foredeck until their surrender was complete, when we would transfer prisoners to the 'D' boats and sink the enemy ship by depth charges or by opening the sea cocks.

In March 1944, when intelligence reported that the German invasion was no longer considered imminent, 643 was ordered to Manfredonia for a short rest and refit. I was told that I was to meet my old friend Tom Fuller and give him a detailed appraisal of the situation in Vis and on the Dalmatian coast. He would shortly take over from me as Senior Officer of the 'D' boats and was awaiting the arrival of a new flotilla, which he would command. He had recently been promoted to the rank of lieutenant-commander and had been awarded the DSC.

Manfredonia was a pleasant small fishing town which was now being developed into a forward naval port. It was good to see Tom again and to be able to swap yarns with him. He seemed particularly interested in my sojourn in Capri. 'Gee,' he said, 'I'd just love to visit that little island,' and between us we soon cooked up a scheme to get us there and back. The Manfredonia naval base was pretty austere, being very short of furniture, especially in the wardroom, which lacked any comfortable chairs. Tom and I saw the commanding officer and assured him that if we could borrow a covered five-ton truck for a few days we would bring him back enough armchairs to equip his wardroom decently. We must have sounded convincing, because the next morning found us under way in a practically new naval truck, bound for Capri via Naples, laughing our heads off at the success of our ruse. We had no idea how we could requisition furniture on Capri, let alone how we could transport it back to Naples, where we would have to leave the truck.

The roads were crowded with military traffic of all kinds, so progress was slow and became even slower as we made the tortuous ascent of the Apennine range, the mountainous backbone of Italy. Our route took us up to about four thousand feet, where great patches of snow were still lying. We were frozen in our lightweight uniforms and getting damned hungry as well. We had brought no rations with us, being confident that the army would look after two itinerant naval officers, but we could find no help anywhere. However, our hopes soared when in the early afternoon we spotted a queue of American GIs formed up outside a canteen in a small town. It didn't take us long to join the queue, but it was half an hour before we got near the serving counter, only to discover that the canteen had just closed amidst groans of dismay and shouts of fury. We then found that the American troops had already had their lunch and that this particular queue was for a special treat – ice-cream! The only comparable thing I could imagine was what would have ensued if British matelots had suddenly been told that the rum issue had run out.

Neither of us was an experienced truck driver and we didn't enjoy the slithery descent down the other side of the mountains, first in sleet and then in incessant heavy rain. When we finally pulled up in the pitch dark in Caserta about 8pm we had had enough. Still thirty miles short of Naples, we located the town major and asked for help in finding food and lodging. He was singularly uncooperative, but when we turned on the pressure and mentioned that we were from the Dominions of Canada and New Zealand he said, 'There are two big military hospitals in the town – a New Zealand one and a Canadian one. You'd better get in touch with them.' Tom and I flipped a coin to see which hospital we should contact first. Tom won and was soon on the town major's telephone. The answer couldn't have been warmer. 'Sure. We'd love to look after you navy boys,' and that they certainly did.

The hospital sent a guide car to show us the way and we fairly basked in the warmth of the welcome as we took drinks and sandwiches with medicos and nurses in a staff lounge. They laid on a special meal for us as well and tucked us up for the night in a light-casualty officers' ward.

Ischia.

It was sheer bliss. In the morning I was wakened by a pretty young nurse bearing a cup of tea to my bedside. Tom was still fast asleep in the next bed. She pointed to his swarthy face and black beard and whispered, 'Does he speak English?' I shook my head. 'Italian?' Another head shake. 'What then?' When I replied, 'Only Canadian,' I think I was lucky not to get my face slapped.

The rain had stopped, the sun shone brightly and all seemed well with the world as we lurched on to Naples in our trusty truck. But the news we were given at naval headquarters was bad. Capri was forbidden to the British and was now a rest and leave resort reserved exclusively for US Army Air Corps personnel. Our hopes plummeted, but were soon raised again when we found out that the island of Ischia was similarly reserved for the Royal Navy. Both islands were long-established holiday and tourist resorts and abounded with hotels and boarding houses. To our simple one-track minds this meant that they must abound with armchairs and other chattels. Within a couple of hours we were being shown into the office of the Naval Officer in Charge, Ischia, only to encounter another set-back. By the greatest ill chance the NOIC proved to be a man with whom I had crossed swords in the past. He disliked me intensely and the feeling was mutual. He was caustic in his refusal to help us, saying that he had never heard of such a preposterous suggestion. He grudgingly agreed to provide accommodation for us – he was in no position to refuse – but he warned us to watch our step as he would be keeping an eye on us.

We were then sent on to another officer, who turned out to be an old acquaintance, Lieutenant-Commander 'Farmer' Lloyd. Farmer, who was one of the most refreshing characters in Coastal Forces, was on Ischia while the 31st ML Flotilla, of which he was Senior Officer, was in for a rest and refit. His specially armed 'B' type MLs made many daring forays in the Adriatic. At the time of our meeting he was convalescing from malaria followed by hepatitis and was making himself useful by helping with the running of the base. Although a fine seaman, he always affected to talk like a stage Somerset yokel – to everyone's delight. Inevitably his officers and ratings proudly called themselves 'Farmer's Boys'. They also indulged in loud choruses from their Flotilla Song, which was that pig-snorting barnyard of a ditty, 'To be a Farmer's Boy'.

Farmer couldn't do enough for us. He too hated NOIC's guts and gleefully helped us to hatch The Great Armchair Conspiracy. During the night he arranged for several splendid armchairs, occasional tables and other furniture to be loaded into a diesel-powered schooner; by dawn Tom and I were half-way to Naples in our commandeered ship. Farmer's parting words had been, 'Just abandon it at some quayside in Naples. No one here will know what in the hell has happened to it.' He added with obvious relish, 'With a bit of luck, its disappearance could get His Nibs into trouble.'

Once at Naples we lost no time unloading our spoils on

Lieutenant-Commander
J I Lloyd

to the truck and made ourselves scarce before anyone could connect us with the stolen ship and furniture. We had had the foresight to refuel before going across to Ischia, so we were able to make a clean getaway. In fact our only stop that day was at the Canadian Military Hospital, partly to thank them again and partly to fuel up with packed sandwiches. One starvation traverse of the Apennines was enough.

At the Manfredonia base, the CO could hardly believe his eyes as armchair after armchair was unloaded and carried into the wardroom. Tom and I solemnly saluted and reported, 'Mission accomplished, sir.' Then all three of us burst into laughter and repaired to the wardroom for a drink. The CO was curious to know where we had been, but we refused to tell him and said that it would be better if he remained in ignorance so that he could never be implicated. Thereafter, whenever Tom or I visited Manfredonia we were given VIP treatment at the base.

April was a great month for Coastal Forces in Yugoslavia. We sank and captured so many enemy ships with their vital cargoes as to constitute a major setback for the German cause in that area. Furthermore, our successes at sea forced the Germans finally to abandon the invasion of Vis and the annihilation of us all which they had so confidently planned only a few weeks before.

Many factors contributed to this splendid result – principally the considerable increase in the number of 'D' boats operating from Vis and the arrival of adequate supplies of a new and vastly improved type of torpedo warhead pistol, which could be set so that the torpedo penetrated a ship before exploding instead of going off on impact. But what transcended all else was undoubtedly the inspired leadership of Tom Fuller, right from the moment he took operational command of the 'D' boats at the beginning of April. No one could have had a more unconventional outlook, and he was full of the most original and ingenious ideas. The first thing he did was to build on the successful boarding techniques we had developed. 'Why sink after capture?' he argued. 'Whenever possible we'll bring 'em back to Vis, complete with cargoes and crews.' Ship after ship was brought to the island and, after its valuable cargo had been off-loaded, handed over to the partisans, to their great delight. The Yugoslavs carried out any necessary repairs, added extra guns and used the vessels further to harry the Germans at sea or for forays against garrisons ashore. Sometimes we had to tow the prizes a matter of fifty or sixty miles.

Tom also felt that there were too many people getting killed or hurt and too many boats suffering damage, and that we had to dream up an entirely new method of attack. Accordingly we began to carry German and Italian speaking partisans with us on patrols. Thereafter schooners were often captured without a shot being fired by either side, their crews having been threatened, in their own language through powerful electrical loud-hailers, with immediate death unless they surrendered at once. We didn't try it on, though, with the deadly German escort gunboats, which were armed with 88mm guns and 40mm cannon. Our torpedoes equipped with the new pistols could penetrate their thick concrete armoured sides, so it was now possible to attack these converted Siebel ferries with a reasonable chance of sinking them. My own Senior Officer, Basil Bourne, arrived with still further reinforcements of 'D' boats and they added to the destruction of German ships, communications and supplies.

In the latter half of May 1944, Tom Fuller departed for a long leave in his home town of Ottawa, with honours thick upon him. Since the beginning of April two more 'immediate' DSCs had been bestowed upon him for his outstanding feats of courage and leadership. We secretly whipped off a telegram to the Mayor of Ottawa:

LIEUT COMMANDER TOM FULLER RCNVR FAMOUS MTB ACE AND TRIPLE DSC NOW RETURNING TO HIS NATIVE OTTAWA ON LEAVE STOP SUGGEST YOU ARRANGE CIVIC RECEPTION STOP CONSULT CANADIAN NAVAL AUTHORITIES FOR DATE AND TIME ARRIVAL OTTAWA STOP

We thought that this rather flamboyant cable would appeal to the North American temperament and to the local patriotism and pride in Ottawa. Anyway, it did the trick. As Tom's train pulled into Ottawa station it was greeted by a cheering throng, a military band, the mayor and other dignitaries, and he wondered what VIP could be aboard. He was more than surprised to discover that the great man was none other than himself.

About the middle of May I received the good news that the Admiralty required me to be routed to England as soon as possible for appointment to 'a more responsible command'. This was a tremendous thrill but very unsettling. With the exciting prospect of soon being able to see my wife and son again, and frequent speculation on what and where my new command would be, I was very much on edge, and I was thankful when I was ordered to Manfredonia to hand over command of my good old 643 – a fine ship that had brought me safely through so many adventures. I'd looked after her and she'd looked after me. I was pleased that by a fortunate coincidence her new captain was Lieutenant Bird, a fellow countryman.

The doctor at the base seemed to think that I must be suffering from nervous hypertension although I personally thought I was fighting fit. In any case there was a week or two to spare before I need to go on to Algiers to pick up the next fast troopship convoy for England. Arrangements were made to send me to an army convalescent home, which was beautifully situated right up in the hills about twenty miles inland from Bari. It was a delightful picturesque village, which must once have been a country resort for the wealthy. There I was given a most pleasant villa entirely to myself, complete with an excellent Austrian prisoner servant. It seemed too good to be true after the sort of life I had been leading, but I revelled in it. The establishment was for officers only. Most of them were recovering from wounds received at Anzio and elsewhere and were members of Britain's more élite units, such as the Guards, Long Range Desert Group, King's Royal Rifle Corps and the Queen's. During my stay there, Basil Bourne and many other old friends from Vis came to wish me well for the future.

I had to take passage in a merchant ship from Bari on 6 June, so I went to the mess early that morning to say goodbye to my army friends who had been so good to me. While we were all breakfasting together reports of the Allied landings in Normandy came through on the radio. It was surely the most welcome news since the war began and I thought it would put everyone in an exultant mood. But no. It was greeted with jeers and the comments were caustic. The general tenor of the remarks was: 'About bloody time those loafing buggers did a little bit of fighting.' The men who expressed these sentiments had seen their regiments decimated in the course of nearly four years of front line action in North Africa, Sicily and Italy.

Rarely can so insignificant a number of small ships have played such a decisive role as they did in Yugoslavia in 1944, when no fewer than ninety-eight German ships were sunk or captured in those waters by British Coastal Forces. Tito was a soldier, not a sailor, but he never forgot how much he owed the Royal Navy for his preservation and for the ultimate defeat of the Germans in Yugoslavia. I believe that when he paid a state visit to England in 1953, he insisted that he would have no guard of honour anywhere unless it was provided by the Royal Navy.

It is interesting to note that a naval engagement, the battle of Lissa, was fought not far from the island of Vis in 1811 and that the victorious British had made a cemetery there in which to bury and commemorate their dead. We thought it only fitting that our own dead should rest beside their countrymen who had fought against another dictator more than a century earlier. Before he left the island Brigadier Tom Churchill, who led our commandos there, had a memorial erected on which are inscribed these words:

AFTER MORE THAN ONE HUNDRED YEARS
BRITISH SOLDIERS AND SAILORS
WHO FOUGHT AND DIED FOR THEIR COUNTRY'S
HONOUR
ON THE SEAS AND ISLANDS OF DALMATIA
HAVE AGAIN BEEN LAID TO REST
IN THIS ISLAND CEMETERY
1944

"They shall be mine, saith the Lord of Hosts,
in that day when I make up my jewels."

13

The distances in the great Mediterranean Sea never ceased to surprise me. It took us just over four days to steam the 1,200 miles from Bari to Algiers. I was the only passenger in this 10,000-ton liberty ship and was most comfortably housed in the comparative luxury and spaciousness of a Merchant Navy officer's cabin. Everyone was very friendly and the trip amounted to another holiday for me. We lived like fighting cocks. During the war, the Merchant Navy was immeasurably better fed than any of the fighting services. There appeared to be no restrictions as far as they were concerned and I was staggered by the wide variety and high quality of their food. Although British, this ship was on the USA run and lacked for nothing. I was shown over the various victualling stores compartments, veritable Aladdin's caves of epicurean delights. Even the frozen meat needed no preparation for the galley. The prime joints, steaks, chops and so on were ready for use – cut, trimmed and individually packed.

I was fortunate in finding some old acquaintances on the staff of naval headquarters in Algiers when I reported there for accommodation and onward routeing. 'Your chances of getting home to England are mighty slim,' they said, 'unless we can keep you out of sight until your ship sails.' They went on to explain that 'press gangs' were about, scouring the ports in search of any naval officers they could find not currently in a sea-going appointment. The press gangs were grabbing anyone they could to man port clearance and operating parties in anticipation of a rapid advance in Italy following the fall of Rome on 4 June. The help of a young surgeon lieutenant was enlisted, and within half an hour I was whisked away in an ambulance bound for an army convalescent resort some forty miles west of Algiers. And so my interesting and variegated holiday continued, this time at the seaside.

A few days later the ambulance returned and I was soon

being carried up the gangway of the *Orion* on a stretcher. Our ambulance and stretcher act was a great success. Not only did it keep me out of the clutches of the press gang but it also ensured me the allocation of a more favourable berth. The *Orion* was then the latest and largest liner in the splendid Orient Line fleet, a 30,000-tonner converted to carry perhaps eight hundred to a thousand officers and men, packed in like sardines. The horrors of such over-crowding have to be experienced to be believed. Even the holds at all levels were filled with troops. The serving of meals went on almost continuously to cope with weary never-ending queues. The sanitary arrangements were likewise completely over-taxed and fresh water was only available for a few limited periods each day. As a result of our stretcher stratagem, I was able to enjoy the luxury of a two-berth cabin with five other officers. We arranged a rota so that each of us had the use of a bunk for a total of eight hours out of each twenty-four.

The 20-knot troopship convoy sailed on 26 June and headed straight for Liverpool. There was no mucking about with course alterations or zigzags. Our high speed made submarine attack difficult and we were protected by the most powerful convoy escort I had ever seen, spearheaded by six or seven modern first line fleet destroyers deployed in an inverted V formation. Two 'Woolworth' carriers with their naval aircraft provided ample air cover throughout daylight hours, not only just above us but for many miles around.

There must have been many thousands of troops in that convoy, culled from all over the Mediterranean to reinforce our armies in Normandy. It was not to be wondered at that, in spite of good organisation and the provision of many special troop trains at Liverpool, it took a long time to clear such a large number of men, together with all their gear and equipment. It was nearly thirty-six hours before my turn came and I boarded a night train for London. We were delayed a few times on the outskirts of the capital because of track damage caused by Hitler's new secret weapon, the macabre pilotless aircraft, the V1, which was literally what it came to be called, a flying bomb. However, by about 11am I reached Sutton Girls' High School, where Helen was still teaching. Helen had no knowledge that I would be in England because I had purposely not written to her in case arrangements had to be cancelled at the last minute, bringing disappointment that would have been hard to bear.

We met in the middle of the playground, which was full of teenage girls. I think our greeting was just 'Hallo, dear.' Unexpectedly we had both become embarrassed and tongue-tied. I walked back with Helen to the room where she had been writing reports. When she courteously asked me to sit down and said, 'I won't be long finishing these papers,' I suddenly exploded and shouted, 'Good God, woman, you're not finishing those damned reports. You're coming out to lunch with me, now.'

It was the first week of July 1944 and all the time we could hear explosions from the V1s. Occasionally we saw one as its motor stopped and it plummeted earthwards. Nothing except dive bombing could be so unnerving. My home leave was short but very sweet. Much sooner than I had expected, I was recalled to active duty.

The situation in Normandy wasn't good. Since the successful invasion landings some five weeks previously, Montgomery had only been able to advance about twenty miles along the coast to the eastward and was held up at Caen by stubborn German resistance. This impasse helped to generate large-scale German naval activity in the Channel. Nearly every night E-boats and the larger R-boats laid mines or made torpedo attacks on the supply routes from England to the great artificial Mulberry harbour at Arromanches, which was still the only usable Allied port in the whole of Northern Europe. The British invasion forces were utterly dependent upon it. On most nights the Germans were running their own convoys along the French coast under the cover of shore batteries; these were also protected by strong escort forces of E-boats and modern destroyers.

Never had there been greater need of experienced Coas-

HMS *Stayner*, Captain class frigate.

tal Force officers and I was appointed as directing officer to HMS *Stayner*, one of the new directing frigates, which were fully manned, each with its own permanent commanding officer and crew. When a directing officer came on board, he in effect took command of the frigate as well as the MTBS and MGBS teamed with her. Once again I became a nocturnal animal, active by night and holed-up by day. My earth was HMS *Dolphin*, the comfortable submarine shore depot, built right at the entrance to Portsmouth Harbour.

Our directing frigates were equipped for locating enemy ships with great precision and for vectoring attacks on them by their accompanying MTBS and MGBS. The scheme was eminently successful but the cost to Coastal Forces was considerable. Because of the efficiency of the directing frigates, our small craft were hurled into fierce engagements with the enemy nearly every night, and the toll of killed and wounded men and sunken or damaged MTBS and MGBS soon mounted to a grievous figure. Sometimes when the radio telephone channel between the frigate and a leading MTB or MGB had been left open, we could hear most of the action as it took place, the gunfire, the shouted orders and even the agonised cries of the wounded and dying – all coming over live as it happened. It was horrible and I found it hard to bear. To be sitting in the comparative safety of the Ops Room of the frigate while directing others into battle was a very different matter from being with them and leading them into an attack.

I decided that my proper place was back in the 'D' boats and that before making the move I must take a course of instruction to bring me up-to-date with the latest developments in German E-boats and in our Coastal Forces in this theatre of war, and also with the tactics being employed. The new 'D' boats were themselves a good example of the tremendous progress being made. Although these formidable vessels were designated as MTBS, they were actually splendid hybrids, combining most effectively the functions of MGBS and MTBS. Their firepower was vastly superior to that of any previous motor gunboat and their four torpedo tubes (as against the two in previous MTBS) doubled their destructive capabilities in that area.

Within three weeks of being appointed as a directing officer, I was sent at my own request to HMS *Bee*, the highly efficient working-up Coastal Force base at Holyhead, for an updating training course.

Before leaving Portsmouth I managed to achieve something which I thought was very worthwhile. I had convinced Captain McLaughlin, who was Captain Coastal Forces, Channel, that each of the directing frigates should carry a doctor and additional sick-berth ratings, to render skilled treatment to the wounded immediately after an action and thus avoid the long delay in getting them back to port first. In my opinion this was the least we could do to help those who fought the battles.

My experiences as a directing officer had taught me much and I had hoped to learn more at HMS *Bee*. But history repeated itself and, just as had occurred at HMS

Seahawk some two years previously, I was again appointed to a command before I could complete the course. My new ship, MTB 725, one of the latest type, was based at Felixstowe and in full operational commission, manned by a crew of experienced officers and men. All these circumstances pleased me immensely, not least because, for the foreseeable future, I would be working in waters that I knew like the back of my hand. Taking over another ship and crew is always difficult and it was a week or two before the ship's company, including the officers, settled for the fact that I wanted MTB 725 smartened up, discipline improved and things done my way, and that I knew what I was about. It wasn't long before we had shaken down together, and everyone was happier for a tighter and better organised ship. We operated on a patrol basis from Felixstowe instead of being controlled by directing frigates.

Nothing exciting happened during my first few trips in 725. But on the night of 17 September, together with three other 'D' boats, we were sent on a mysterious mission to a specified zone near Dunkirk, which was still held by the Germans. There we were told we would encounter some German ships. This was obviously a tip-off from naval intelligence, but we found nothing either that night or the next, when we were again sent there. On the third night we split our force to increase our chance of making an interception, sending two of our boats to patrol ten miles offshore, while 725 and her consort, commanded by 'Biscuits' Strang, a delightful Glaswegian who was half-leader of our flotilla, patrolled close inshore. Just after 11pm we picked up some garbled German wireless orders and soon afterwards overheard somewhat confused and confusing signals which one of our offshore patrol boats was making to base. We gathered that the two offshore 'D's had encountered three E-boats and that in the ensuing action two of the E-boats and one of the 'D's had been sunk. The 'D' boat returning to base had casualties but there was no mention of picking up any survivors.

It was unthinkable to us that either friend or foe should be left adrift and abandoned in the cold North Sea, so we decided to mount a search with our consort. We thought there were bound to be some survivors in rubber rafts or life-jackets. We ploughed through a September mist which soon became a fog, reducing visibility to little more than fifty yards. After a few hours in such adverse conditions we began to feel that our efforts were pretty hopeless, but although we were ordered back to base we ignored the signal and persevered in the hope that full daylight might disperse the fog. This didn't happen, but after eleven pains-taking, monotonous hours of peering through the murk, our stubborn endurance was gloriously rewarded when we sighted rubber rafts piled up with German sailors, many of them wounded. In the water nearby, wearing a life-jacket, was a very smartly dressed officer, with his binoculars still slung round his neck.

I knew that if we attempted to go alongside those rafts one touch from our ship would be sufficient to capsize them. 'Can you speak English?' I shouted to the German officer. 'Fluently,' came the reply. 'Listen carefully then. I'm sending you a heaving line and I want you to make it fast to each raft in turn so that we can haul them along-side.' This he did, and we lowered our scramble nets. Some of our officers and men went up to their shoulders in the water to help the men on the rafts on board, handling the wounded with infinite care. In all we saved more than fifty of them. It was now a matter of urgency to return to base so that the wounded could have proper medical attention. We were ordered into Harwich and on the way our crew did all they could to see to the comfort of our unexpected guests – hot drinks, sandwiches, blankets, towels, drying out of clothes – the lot. Even those who were not hurt were suffering from exposure and shock, except the Ger-man officer, who was none other than Kapitänleutnant Karl Müller, holder of the Knight's Cross and one of the greatest of the E-boat leaders. He had just fought his 164th action. What a man – as tough and yet as chivalrous as they make them.

It was he who had remained in the water all those hours rather than overload any of the rafts further. After a rub-

S 112, which Karl Müller commanded in 1942. Armed with four torpedoes, a 40 mm gun and two 20 mm guns, she had a maximum speed of 43 knots.

down with a towel and a change of clothes – my flannel trousers and sports jacket fitted him very well – he seemed fully recovered and showed no sign of the ordeal, both by fire and by water, that he had been through. When he saluted me and formally introduced himself, I realised I had caught a very big fish indeed; but I didn't know that our meeting was to be the start of a deep friendship that endures to this day. Since the war Karl and I have met frequently both in England and Germany, staying in each other's homes, and more recently exchanging visits between Germany and New Zealand.[1]

Karl was profuse in his thanks for the rescue we had made and handed me his German naval Zeiss binoculars as a gift, in recognition of our having saved the lives of

[1] Author's note. On a visit to Germany in 1978 I had the privilege of being invited by Vice-Admiral Klose, the Commander-in-Chief of the German Navy, to meet him at his headquarters at Flensburg. He had been a contemporary of Karl's and like him had commanded an E-boat flotilla. It gave me a very strange feeling to be in the building from which German naval operations had been directed during the dark days of 1939–1945.

him and his crew. I still have these magnificent glasses and have used them constantly ever since, particularly when cruising in my yacht.

On arrival in Harwich the wounded were dealt with at once, but it was some time before an armed guard came on board to take our other prisoners away, blindfolded. While we were waiting, Karl and I yarned together in my cabin. 'I've got a little present for you,' I said, handing him a bottle of whisky. He looked at me in surprise, not quite sure what he was supposed to do with it. But when I produced two glasses and said gently, 'Aren't you going to offer me a drink?' a huge smile spread over his face and he poured out two stiff ones. We just managed to finish the bottle before the guard detachment arrived and led him away. We were both in good heart by then and must have looked like a couple of cats that had swallowed the cream.

As soon as we had been cleared of our German prisoners, we steamed across the harbour to our own base at Felixstowe, where we were glad to learn that both of our offshore 'D' boats had returned safely, although not without some damage. This tended to confirm the account of the action that Karl had given me. He had told me that when he sighted the two British boats he had turned sharply away, but unfortunately his two following E-boats collided in the fog. They were in a sinking condition by the time he had located them. Karl was carrying out rescue operations when the two 'D' boats found them and raked them mercilessly with cannon and machine guns, destroying his ship as well. Shortly after the British boats had disengaged they too became victims of the fog and, mistaking each other for an enemy, fought a brief but lively action between themselves during which one of them caught fire. Here obviously was the explanation of the confused signal from the 'D' boat to base which we had overheard earlier, reporting that two E-boats and one 'D' had been sunk. It was clear also that, sadly, the British casualties (three dead and one wounded) had been self-inflicted.

Soon after the war, when reading an important new book about Coastal Forces, I was irritated to come across

Kapitänleutnant Karl Müller in 1943.

an erroneous description of this action. It gave the date as being 18 September, whereas it was actually the night of 19–20 September and stated that the affray was 'perhaps the most decisive battle against E-boats of the whole war'.[1] Mention is made of the British boats shelling each other and also that 'over sixty prisoners were taken, amongst them Müller himself'. But it reads as though the Germans had been rescued immediately after the action by the same boats that had been engaged in it, and not, as actually happened, by another ship altogether (725) eleven hours later. I am in no doubt that Karl Müller's version is the correct one and do not think therefore that it was quite the victory it was made out to be at the time, particularly as we were to learn later that our two MTBS were supported by the powerful gunfire of HMS *Stayner* in their attack on the already sinking E-boats.

Karl Müller was regarded as a prisoner of such importance that he was taken over to Supreme Allied Headquarters in France where he was interrogated for two days. When we met again in Germany after the war, he told me about this and about some of his subsequent adventures. He had been greatly impressed by the efficiency of British naval intelligence. Apparently they had a complete dossier on him, which not only covered his naval training and career but also included information about his parents and even his wife's parents. An interesting personal detail was noted to the effect that Müller was a confirmed anglophile and was frequently called 'Charlie' in the German navy. He showed me the large circular silver tray given to him as a wedding present in 1942 and inscribed with the signatures of every officer in his flotilla. In the centre were the words 'To our Charlie with the best wishes of us all'.

I was interested to learn also that the tip-off naval intelligence had given us had been absolutely correct. Karl said that he had planned to be in the zone we were patrolling on 17 and 18 September to escort some ammunition ships to the beleaguered German garrison in Dunkirk, but their

[1] *The Battle of the Narrow Seas* by Lieutenant-Commander Peter Scott (London, 1945) p.215

sailing was delayed to await the misty conditions that the German meteorological experts forecast would soon occur, as they did on 19 September when Karl was able to run his convoy through to Dunkirk safely and undetected because of the poor visibility. It was a fickle turn of fate that this fog, which had been so favourable for his plans at first, should have been so disastrous for him later. But luck hadn't entirely deserted him.

Almost immediately after his interrogation, Karl was released and sent back to Germany in exchange for a British officer who had been wounded.[1] He little knew that the most interesting period of his naval career was yet to come. He was appointed to the personal staff of Admiral Dönitz, Commander-in-Chief of the German Navy, and thereby came in contact, directly or indirectly, with the most important people in Germany, including Hitler himself. As a result of the high mutual regard that developed between Dönitz and Karl, their friendship was maintained after the war and indeed right up to the admiral's death in 1982. In his will Dönitz paid a final tribute to Karl by nominating him to be one of his pall bearers for his last earthly journey.

14

Two or three days after the Müller incident I was ordered to Lowestoft from where I was to mount operations on the Dutch coast with MTB 725 and a sister ship, acting as Senior Officer of this small unit. My memories of Lowestoft were not happy ones and I was dismayed to find that the

[1] The historian M R D Foot, then a captain in the SAS, who had been captured in southern Britanny on 24 August 1944 while on a secret mission. By remarkable good fortune a souvenir-hunting guard removed his SAS wings, leaving him dressed as a captain in airborne Royal Artillery. His interrogators, not realising that he spoke German, assumed that he was on reconnaissance from the 6th Airborne Division. After his fourth attempted escape on 20 November, when he was badly wounded, Captain Foot was included in an exchange of prisoners of war negotiated by an American Red Cross official. He did not know the identity of his German opposite number, except that he was a highly decorated officer.

commanding officer of the Coastal Force base there was the same one who had threatened to court-martial me on my last visit in 1941. He didn't strike me as being the type of man who could forgive and forget, and when I reported to him he made it quite clear that he had done neither. It was obvious that I was going to be in for a hard time for as long as I had to be under his command.

However, Great Yarmouth was not far away and my affection for that base was as deep as my repugnance for the Lowestoft one. A visit there would surely be the best way to dispel the mood of despondency that had fallen upon me; within an hour I strode jauntily into Shaddingfield Lodge, a splendid sea-front property which had been requisitioned to serve as an officers' mess for Coastal Forces in Great Yarmouth. The Lodge had an interesting history. In the early years of this century it had been acquired by the then Prince of Wales to house his mistress, the world-famed beauty Lily Langtry.

When I arrived at Shaddingfield I was disappointed to find no one else about. I sat down and waited for events to shape themselves. It was not long before two familiar figures came through the swing doors and made for the bar. One was Commander Brind, who had been in charge of the Great Yarmouth base during the time I was in the 1st Flotilla, and the other was Captain Robson, whom I had served with when he was a staff officer with Admiral McGrigor in Sicily and Italy. It was a happy coincidence that we should all walk into that mess at practically the same time on that particular day. I can still recall both the scene and the conversation quite clearly.

Brind was so startled when I appeared beside him at the bar that he ejaculated, 'Good God, Hobday, where on earth did you spring from?' and tapping Robson on the shoulder said, 'This is the man I've just been telling you about, the one who took his ML out on an anti-E-boat patrol on hand steering when we were short of boats.' Robson had turned to Brind when he heard my name. 'Hobday and I are old friends,' he said. 'I came to know him well in the Mediterranean and I share your opinion of him.' I found this remark rather embarrassing, but as they grinned at each other like Cheshire cats and asked me to join them I assumed it was genuinely meant. I learned that Captain Robson was now Captain Coastal Forces, East Coast and that Commander Brind was on a few days' leave from the new Coastal Force base he had been setting up at Ostend.

Many old acquaintances looked in over the next half hour and greeted me with considerable surprise. It was good to meet them all again but what gave me most pleasure was to see my staunch old friend Bob Harrop. Our last encounter had been when he was second-in-command of my HMS *Seahawk* anti-submarine class at Ardrishaig. After Commander Brind's departure Captain Robson invited me to his quarters for dinner.

As soon as I was seated in his spacious apartments, he pulled up a chair and without any preamble shot out, 'I'm surprised to see that you are still in the rank of lieutenant and that you aren't wearing a decoration. Something must have gone terribly wrong for you and I want to know all about it. And don't try to hold anything back from me.' I had no alternative but to tell him all about my experiences in the Dodecanese and in Yugoslavia and to bring him up-to-date with details of my appointment to the directing frigate and then to MTB 725, finally mentioning my difficult position at Lowestoft. He listened carefully and asked a number of pertinent questions. Then he said quietly, 'I think that you have been most shabbily treated and I'm going to do something about it right away; but a recommendation for your promotion must be initiated by your immediate commanding officer, which means your friend the commander at Lowestoft.' He smiled as he added, 'I don't expect he is going to like that but I think I can persuade him.' Robson pushed a bell and asked his secretary to get Commander Lowestoft on the line. My ears burned as I overheard the conversation, which finished up with, 'Hobday is staying the night here as my guest and I am sending him back to Lowestoft in my car to report to you at ten o'clock in the morning.' When Robson had hung up he said, 'As you may have gathered, your promotion is all arranged.' I was overwhelmed with elation and will always remember Geoffrey Robson's kindness to me. To round off my stay at Shaddingfield, I had the use of 'The Jersey Lily's' bathroom in the morning. It had been carefully preserved and was a fascinating room full of wall mirrors. The taps were gold plated and the bath was egg-shell blue enamel.

My promotion came through very quickly and the leopard did change his spots. The commander at Lowestoft now exhibited a most benevolent and friendly attitude towards me. My new rank of lieutenant-commander was a rare enough distinction among temporary RNVR officers to warrant a degree of self-consciousness in the early days of promotion, but the first time I wore my extra ring I made a fool of myself. I was deep in thought about the operation we were to carry out that night and was surprised that the sentry should present arms instead of making the usual butt salute as I entered the base. Thinking there must be some high-ranking officer immediately behind me, I quickly turned round to have a look. There was no one there and I realised that the present was for me. The sentry must have thought that I was quite mad.

My first lieutenant in MTB 725, the Hon Euan Howard,[1] and Peter Vanneck,[2] the CO of MTB 696, my other 'D' boat, were both very well connected young gentlemen. Euan was a son of Lord Strathcona and his grandfather had been Chairman of Canadian Pacific, which used to advertise itself, quite truthfully in view of its vast ownership of railways, shipping and hotels, as 'the world's largest transportation system'. Peter was also related to peers of the realm. One leave night he suggested that we went and had a bite with his uncle and aunt, Lord and Lady Huntingfield, who had an estate in Suffolk only about twenty miles away. Euan and Peter both had motorbikes and I rode pillion with Peter. Eventually we turned in through some lodge gates and followed the mile-long drive to the front

[1] Later Lord Strathcona and Mount Royal (b.1923). Minister of State, Ministry of Defence, 1979–1981.

[2] Later Air Commodore the Hon Sir Peter Vanneck (b.1922). Lord Mayor of London, 1977–1978.

door of Heveningham Hall, one of the largest country houses I have ever seen. Peter tugged on the bell lanyard and after a long wait an elderly lady appeared. It was his aunt. We entered the magnificent hall, two or three storeys high and pure Adam. We followed Peter's example in putting our caps on the heads of various marble busts, and then walked through the long gallery, hung with old masters, to the butler's quarters in the east wing. The Long Gallery was aptly named and I could see why it had been some time before Peter's aunt had come to the door.

In the butler's small dining room we met Peter's uncle and two distinguished visitors. They were General Jimmy Doolittle, commander of the US Eighth Air Force, and his staff colonel, delightful people both of them, as were indeed the uncle and aunt. They had just finished dinner. After pouring us a drink our host chivvied the general and his aide into the scullery to do the washing up, saying, 'We've got to look after these boys. They're doing front-line fighting, not living soft as we are!' Living soft in some ways perhaps. But this dear old couple, with all their wealth and used to every luxury, had had to move into their butler's quarters to conserve their fuel ration and because no servants were available during the war. I suspect that they found this camping out quite good fun in the short term.

While the distinguished American guests continued their menial work in the scullery, our host and hostess served us with a splendid meal, the main course of which had been kept hot in the oven. Happily the high-ranking Americans completed their chores at the same time as the low-ranking British finished their delectable repast; and we were able to join together and yarn away while Lord Huntingfield provided us with a magnificent port, which was passed many times. As we left we were each given a brace of pheasant. Each brace was tied together by the legs and as it was frosty we slung them round our necks and tucked them into the front of our monkey jackets. We were a happy trio as we rode back through that sparkling night, kept warm by our pheasant cravats without, and by roast partridge and a great burgundy within.

Messdeck of an MTB.

We hardly seemed to have settled down in Lowestoft before we were ordered to proceed to Portland via Dover. Although I didn't know it at the time, this was to be my last voyage in command of one of His Majesty's ships and it certainly completed my gamut of experience in Coastal Forces. I had commissioned ML 339 and MGB 643 after acceptance trials from their building yards and I had inherited MTB 725 as a going concern. Now I was under orders to pay off 725 and her sister ship at Portland.

We had a rough passage to Dover and by the time we got there it was blowing great guns. The heavy seas were savaging the breakwater and exploding into thunderous gouts of spray. We were thankful to reach smooth water again and berthed at a quay in the north-eastern corner of the submarine harbour, where another 'D' boat was tied up. As soon as we had made fast we were invited to go aboard. It is a usual naval courtesy for a ship already in harbour to send such an invitation to those just come in

Engine room of an MTB.

from sea. The CO of MTB 713 was Lieutenant Dick Olivier, brother of the famous actor Laurence Olivier, but very much a character in his own right. I was interested to meet his first lieutenant too, Sub-Lieutenant Low, a strapping and forthright young New Zealander who looked extremely fit.

Olivier lost no time in telling us how glad he was to see our two ships come sailing in. 'I need all the help I can get from people like you,' he said. 'I've been placed in a damned awkward position through no fault of my own and it's a situation that needs drastic action.' He then related his tale of woe. He had only just fully restocked his ship with duty-free wines and spirits when, out of the blue, he had been ordered to proceed to Portland forthwith to pay off. He couldn't land the stuff or return it without having to pay duty, which he said he couldn't afford. 'So I'm throwing a series of parties on board to consume it all before we get to Portland. What a Godsend this gale is! It should give us another two or three days' grace.'

Naturally we commiserated with the poor fellow in his difficulties and promised to give him all the support we could. But as soon as we returned to our own ships I said to my officers, 'We will keep clear of Olivier until the muster and inventory on both ships is completed ready for pay off.' They agreed. Meanwhile, parties went on day and night aboard Olivier's ship, and it was obvious that the gallant fellow was putting up a courageous and well sustained fight. The gale continued to blow, but on the third day it started to ease off and I decided to make passage to Portland the following morning. As our ships were not on operations there had been no reason to brave the storm. We had got as far as we reasonably could with our musters and inventories, and on the afternoon before we sailed we were quite ready to help ease Olivier's drink problem.

A rip-roaring party was in progress as we entered the crowded wardroom through a haze of cigarette and cigar smoke. Our host seemed to be standing up to the punishment quite well, and welcomed us warmly as he introduced us to the mayor, harbour master and other local dignitaries. After a while Olivier pressed the bell and sent for his first lieutenant, who reported quite curtly, obviously fed-up to the back teeth with his CO and his parties. Olivier turned to him and said in a very casual tone of voice, 'It's frightfully stuffy in here, Number One. Do something about it. We need more ventilation.' I knew we were in for trouble when I saw Number One's face light up with an almost unholy joy and he yelled, 'Bring me the junk-axe, coxswain!' The bulkhead between the main alleyway and the wardroom was made of thin wooden planking. Shouting, 'Watch out inside!' the first lieutenant rapidly chopped out a neat hole, about four feet long and a foot deep. Pieces of wood flew all over the wardroom. In the total silence which followed this impetuous display of good axemanship and bad temper Number One asked, 'Is that better, sir?', no doubt hoping he had goaded his captain into a fury.

The ploy was unsuccessful, and it was Number One himself who was maddened when Olivier remained quite unruffled, replying lazily in his cultured English voice, 'Much better, thank you, dear boy.' The wardroom looked as though a bomb had exploded in it. Chunks of wood lay scattered everywhere and acrid fumes of smoke swirled through the splintered gap in the bulkhead. Soon everyone was picking up pieces of planking and getting them signed by all the others, as sometimes occurs with menus at notable banquets. As each of us departed we proudly bore with us one of these autographed souvenirs of a remarkable occasion.

When we steamed into Portland I was expecting trouble. I had had to take over MTB 725 at short notice and there had been no time to muster and check the massive inventory. I had had to sign the takeover, blind, and put to sea immediately on operations. Our muster at Dover had shown that there were innumerable items missing, some of them quite costly ones. As the commanding officer of one of His Majesty's ships, I had good reason to be worried because I could be held liable for the value of items unaccounted for. But the run of good luck I had enjoyed since my visit to Great Yarmouth still continued. Quite unexpectedly, the paymaster commander controlling the paying-off of our ships turned out to be Neville Rose, a departmental manager in ICI before the war with whom I had had many pleasant dealings. To my great relief he arranged everything to my advantage with practised professional skill, using such terminology as 'broken due to stress of weather', 'destroyed by enemy action' and 'lost overside in heavy weather'. 'You'll have to pay for a few things, Geoff, particularly those which can't be accounted for by being lost overboard etc. Otherwise it will look fishy.'

With the expert cooperation of Neville Rose, the paying-off of MTB 725 was completed with startling rapidity. The next day I was instructed to report to the Admiralty in London to be briefed on my new appointment. On the train, I kept my fingers crossed, hoping against hope that I would now be given command of a frigate or an escort destroyer. Such an appointment still remained my greatest ambition, but this was not to be. To my utter surprise the Admiralty offered me something much better, an opportunity I would never have dreamt of. I was to be flown out to Australia to join the staff of the British Pacific Fleet, which was in the process of formation under the command of Admiral Sir Bruce Fraser. Very conveniently, I would be based in Melbourne, where nearly all my relatives lived.

The British Pacific Fleet was to be the largest single British naval force of the war and to consist of only the most modern and powerful of our fighting ships. My new chief, Commander Kenneth Kemble – a lawyer by profession – was to be the head of the transportation and port facilities staff. He had joined Coastal Forces in 1941 and had been given command of MTB 104 based on Felixstowe. After service in the Mediterranean, he returned to the United Kingdom to become Senior Officer of the 19th ML Flotilla and subsequently Deputy Training Commander of HMS *Bee*. I had met him a few months earlier when I was based at HMS *Dolphin*. Admiral Fraser had selected his staff on an entirely new basis. Half of us were officers like myself who had been on continuous active service at sea throughout the war. He knew we would be strongly motivated to see that the fleet was kept supplied with all its needs regardless of red tape or any other impediments. The other half of the team was made up of experienced 'professional' staff officers to guide us in the best procedures to accomplish our purpose. In practice this novel arrangement worked extremely well and I found the experience exhilarating. My new job which I held until the end of the war in the Far East, turned out to be the most important and constructive one of my entire naval career.

Nevertheless, I was frequently assailed by a deep nostalgia for the incomparable comradeship and adventurous life that I had enjoyed in the dangerous and testing times of the past three and a half years. Serving in Coastal Forces was a privilege which I have always recalled with honest pride; and my memories of those splendid little fighting ships and the gallant men who fought in them are as vivid to me now as if it had all happened yesterday.

Index: People

Editor's note
Ranks and decorations are contemporary with this account.

Index: Units

SEARCHLIG
PORT OUTER ENGINE UNDER
TWIN 20mm OERLIKON GUNS
DEPTH CHARGE
PORT INNER ENGINE UNDER
6 POUNDER
STBD INNER ENGINE
STBD OUTER ENGINE
DINGHY
POS MESS UNDER
SMOKE APPARATUS
FUEL TANKS
STBD OUTER SCREW
STBD RUDDER